CfE Higher
PHYSICS
SUCCESS GUIDE

Michael Murray

001/180515

10 9 8 7 6 5 4 3 2 1

ISBN 9780007554379

Published by
Leckie & Leckie Ltd
An imprint of HarperCollins*Publishers*
Westerhill Road, Bishopbriggs, Glasgow, G64 2QT
T: 0844 576 8126   F: 0844 576 8131
leckieandleckie@harpercollins.co.uk   www.leckieandleckie.co.uk

Special thanks to
Lee Mulvey Haworth (project management); Roda Morrison (copy-edit); QBS (layout and illustration); Nick Forwood (proofreading); Ink Tank (cover design)

A CIP Catalogue record for this book is available from the British Library.

Acknowledgements
Whilst every effort has been made to trace the copyright holders, in cases where this has been unsuccessful, or if any have inadvertently been overlooked, the Publishers would gladly receive any information enabling them to rectify any error or omission at the first opportunity.

Printed in Italy by Grafica Veneta S.p.A.

Unit 1: Our dynamic universe

Unit 2: Particles and waves

Contents

The Higher Success Guide

About this Success Guide

The guide covers the mandatory content in the key areas of the Higher Physics course.

The revision material is laid out over a double page (sometimes three pages) in a clear, concise way. The main headings and sub-headings make it clear which part of the course is being covered. There are 'Top Tips' throughout, which emphasise important points and give hints on how to answer questions. Each section has a quick test to enable you to self-assess so that you know your areas of strength and areas you need to look at in more detail. Answers to each test can be found at the back of the guide.

The Higher Physics Course

Structure

The CfE Higher Physics Course is made up of the following units:

- Our dynamic universe
- Particles and waves
- Electricity
- Researching physics

In order to achieve a pass at Higher, you need to pass all of the units as well as the course assessment.

Course assessment

This takes the form of a question paper (exam) and an assignment.

The exam carries a total of 100 marks and is made up of two sections:

- the objective test (20 marks) which comprises 20 multiple-choice questions
- Paper 2 (80 marks) which is made up of a mixture of restricted and extended-response questions.

The majority of marks will be awarded for applying knowledge and understanding. Other marks will be awarded for applying scientific inquiry, scientific analytical thinking and problem solving skills.

A data booklet containing relevant data and formulae will be provided.

The assignment is a written report, based on your own investigations and research into a topic in the course. It is worth 20 marks.

Preparation for both of these is critical to getting the best grade possible in your Higher Physics Course.

Properties of motion

Scalars and vectors

Quantities can be defined as scalars or vectors. A scalar quantity is defined only by its magnitude, or size. A vector quantity is defined by both its magnitude **and** direction.

Some examples of scalar and vector quantities include:

Scalar quantities	Vector quantities
Distance	Displacement
Speed	Velocity
Time	Acceleration
Mass	Momentum
Energy	Force

Displacement and distance

Distance is the measure of how far an object has travelled. As it is a scalar quantity, it can be described by magnitude **only**.

Displacement measures how far an object moves from its starting position, and the direction in which it has travelled. It requires both magnitude and direction, and so is therefore a vector quantity.

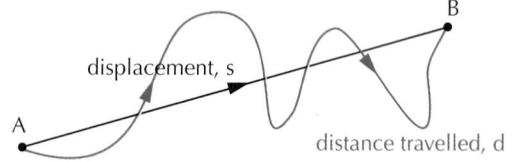

Speed and velocity

Speed is the distance covered in a unit of time. As distance, d, and time, t, are both scalars, so speed is also a scalar quantity. The average speed is given by:

$$\text{Average speed} = \frac{\text{distance}}{\text{time}} \qquad \longrightarrow \qquad \bar{v} = \frac{d}{t}$$

Velocity is the displacement involved in a unit of time. As displacement, s, is a vector, velocity is also a vector, one with both magnitude and direction. The direction of the velocity is in the same direction as the displacement.

$$\text{Average velocity} = \frac{\text{displacement}}{\text{time}} \qquad \longrightarrow \qquad \bar{v} = \frac{s}{t}$$

Acceleration

Acceleration is the rate of change of velocity for an object. Since the change in velocity is measured in m s^{-1} and time is measured in seconds, acceleration is measured in m s^{-2}.

$$a = \frac{v - u}{t}$$

where u is the initial velocity and v is the final velocity.

Acceleration is a vector quantity, and so takes place in a particular direction. If an object speeds up, slows down, or changes direction it is said to be accelerating. We indicate a difference in direction by using a positive or negative sign.

Top Tip

A negative acceleration is the same as a deceleration, i.e., an acceleration of -5 m s^{-2} is equivalent to a deceleration of 5 m s^{-2}.

Adding vectors in two dimensions

When we add vectors together we find the resultant. The resultant is a vector that can replace all vectors present with a single vector. This technique can be used for all vector quantities, e.g., displacement, velocity and acceleration.

Vectors can be drawn with an arrow to indicate the direction. The length of the arrow represents the magnitude. Vectors must then be joined tip-to-tail.

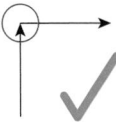 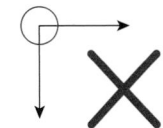

The vectors can be joined in any way, as long as their direction is maintained. Once the vectors are joined together, join the start and end points to find the resultant vector.

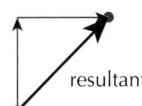

resultant

The resultant must be stated with both its magnitude **and** direction, e.g., 150 m at 37° E of N, or 150 m at 037. If the triangle is right angled, it can be calculated using Pythagoras and trigonometry. If the triangle is not right angled, a scale diagram will be required. Alternatively, the sine and cosine rule can be used.

Quick Test 1

1. Describe the differences between scalar and vector quantities.
2. What is meant by *displacement*?
3. What is meant by *velocity*?
4. What is meant by an acceleration of 2 m s^{-2}?
5. What is meant by the resultant?

Motion-time graphs

Displacement–time graphs

A journey can be represented by a graph of displacement against time. This allows us to determine the displacement of an object at a given point in time. It also allows us to calculate information about the velocity, and therefore the acceleration of the object.

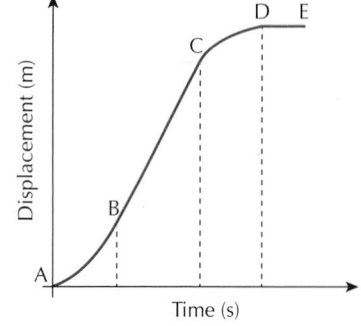

The increasing slope between A and B on the graph shown opposite indicates an increasing velocity (an acceleration). The constant upward slope between B and C represents a constant velocity. The decreasing slope between C and D indicates a decreasing velocity (a deceleration). Finally, the horizontal line between D and E is an indication that the object is stationary with zero velocity.

We can summarise the following for displacement–time graphs:

1. A straight line indicates a constant velocity.

2. The gradient of a displacement–time graph is equal to velocity.

Velocity–time graphs

A journey can be represented by a velocity–time graph. This not only allows the velocity of the object to be determined at a specific moment in time, but also enables the displacement and the acceleration to be calculated.

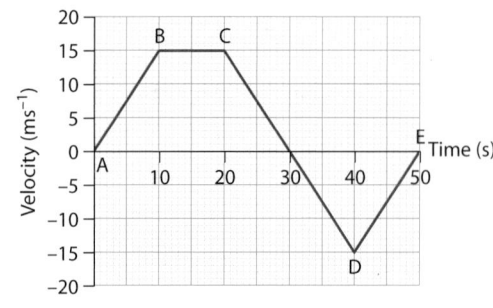

The positive slope between A and B on the graph shown opposite indicates a positive acceleration. The horizontal line (or zero slope) between B and C indicates a constant velocity, and therefore no acceleration. The negative slope between C and D indicates a negative acceleration (a deceleration). The positive slope between D and E indicates a positive acceleration.

We can summarise the following for velocity–time graphs:

1. The displacement is equal to the area under a velocity–time graph.

2. The gradient of a velocity–time graph is equal to acceleration.

TOP TIP

Negative values on the velocity axis indicate a change of direction.

Acceleration–time graphs

An acceleration–time graph can be used to display how the acceleration changes over the course of a journey.

The acceleration–time graph can be produced from the data in a velocity–time graph.

The acceleration–time graph is produced as follows:

$$AB: a = \frac{v-u}{t} = \frac{15-0}{10} = 1.5\,ms^{-2}$$

$$BC: \text{constant speed} \therefore a = 0\,ms^{-2}$$

$$CD: a = \frac{v-u}{t} = \frac{(-15)-15}{20} = -1.5\,ms^{-2}$$

$$DE: a = \frac{v-u}{t} = \frac{0-(-15)}{10} = 1.5\,ms^{-2}$$

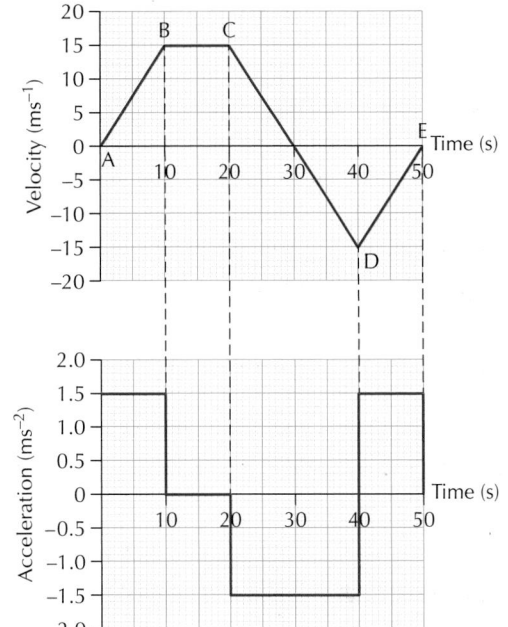

Objects thrown upwards

The velocity of an object thrown upwards can be represented in a velocity–time graph.

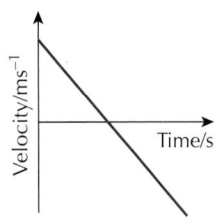

The velocity is positive as the object travels upwards to its maximum height where the velocity is zero. The velocity is negative as it travels downwards. The gradient of the graph is constant, indicating a constant acceleration due to gravity.

> ## Top Tip
> We only deal with constant accelerations in Higher Physics. An acceleration–time graph should therefore only ever include straight horizontal lines.

Bouncing ball

The velocity of a bouncing ball can be shown in a velocity–time graph.

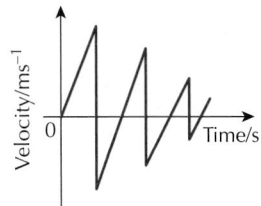

The velocity is positive as the ball travels downwards. When the ball leaves the ground, a negative velocity indicates the change in direction. The velocity of the ball is zero at its maximum height. There is some loss of energy at the ground which means that the ball will not bounce up to the same height.

Quick Test 2

1. What quantity can be calculated from the gradient of a displacement–time graph?
2. How can we calculate displacement from a velocity–time graph?
3. What quantity can be calculated from the gradient of a velocity–time graph?

Equations of motion

Deriving the equations of motion

There are three equations in Higher Physics that we can apply to objects moving in a straight line with constant acceleration. We call these **equations of motion**.

The first of these comes from the equation for acceleration:

$$a = \frac{v-u}{t}$$

$$v - u = at$$

$$\mathbf{v = u + at}$$

The second can be derived from an understanding of graphs of motion. The displacement, s, from the starting position is given by the area under the velocity–time graph.

$$s = area\ of\ 1 + area\ of\ 2$$

$$s = ut + \frac{1}{2}t(v-u)$$

but $\quad v - u = at$

$$\therefore s = ut + \frac{1}{2}t(at)$$

$$\mathbf{s = ut + \frac{1}{2}at^2}$$

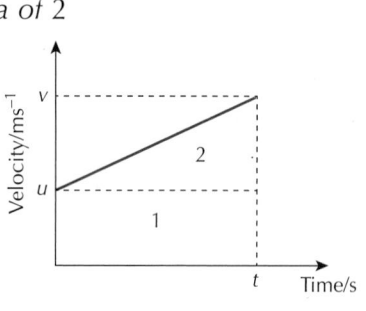

Finally, the third equation of motion can be found by combining equations 1 and 2:

$$v = u + at$$

$$v^2 = (u + at)^2$$

multiplying out the brackets

$$v^2 = u^2 + 2uat + a^2t^2$$

$$v^2 = u^2 + 2a\left(ut + \frac{1}{2}at^2\right)$$

But $\quad s = ut + \frac{1}{2}at^2$

$$\therefore \mathbf{v^2 = u^2 + 2as}$$

TOP TIP

You will not be asked to derive the equations of motion in assessments.

Applying the equations of motion

The equations of motion can be applied to any object that is moving with a constant acceleration in a straight line.

When applying the equations of motion, it is a good idea to list all the possible quantities involved.

Symbol	Quantity	Units
s	displacement	metres (m)
u	initial velocity	metres per second (m s^{-1})
v	final velocity	metres per second (m s^{-1})
a	acceleration	metres per second squared (m s^{-2})
t	time	seconds (s)

An easy way to remember all the quantities is **'suvat'**.

TOP TIP

Start each equations of motion problem by writing down 'suvat', and filling in the quantities you have values for. This makes it much easier to identify which of the equations of motion to then use.

Quick Test 3

1. What is the first step in an equations of motion calculation?

2. A train is moving with a speed of 2 m s^{-1}. The train now accelerates at 4 m s^{-2} until it reaches a speed of 15 m s^{-1}. Calculate the distance travelled by the train during this acceleration.

Balanced and unbalanced forces

Newton's first law

Newton's first law states that:

> **An object will remain at rest, or will continue to move in a straight line at a constant speed, unless acted upon by an unbalanced force.**

For example, during a car crash the occupants will continue to move forward at a constant speed, if not wearing seatbelts, which provide an **unbalanced force** to stop them.

Most moving objects do not continue to move at a constant velocity, due to either the force of friction or the force of gravity.

Newton's second law

Newton's second law states that:

> **A uniform unbalanced force will produce a uniform acceleration.**

This is often expressed in the relationship

$$F = ma$$

It is this law that is used to define the unit of force, the Newton (N). An unbalanced force of 1 Newton will accelerate a mass of 1 kg at 1 m s^{-2}.

TOP TIP

The 'F' in F = ma is the *unbalanced force* in Newtons. This means that if there is more than one force acting, then you must calculate the resultant force.

Example

A car has a mass of 1000 kg. The engine provides a force of 10 000 N. The frictional force acting on the car is 5000 N. Calculate the acceleration of the car.

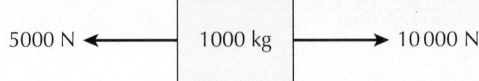

5000 N ← | 1000 kg | → 10 000 N

Unbalanced force = 10 000 − 5000 = 5000 N

$F = ma$

$5000 = 1000a$

$a = 5 \text{ m s}^{-2}$ to the right

Terminal velocity

In reality, the frictional force acting on an object does not remain constant. In fact, friction increases as velocity increases.

Consider the example of a skydiver.

The downwards force acting on the skydiver is weight ($W = mg$), which causes an acceleration towards the ground. The force of air resistance acts against the skydiver. The resultant force acting on the skydiver can be calculated from the difference between these two forces.

As the velocity of the skydiver increases, the air resistance upwards increases. Because the weight force downwards is constant, the difference between these two forces grows smaller. This eventually results in the forces becoming balanced, and the skydiver falls with a constant velocity. This is called the **terminal velocity**, and is the maximum velocity that the skydiver can reach.

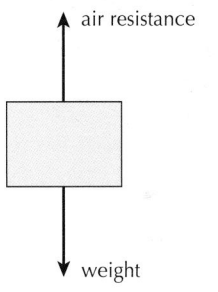

air resistance

weight

Tension

Tension is the force that is transmitted through a cable, rope, string or wire when it is pulled tight by forces acting from opposite ends. Tension acts along the full length of the string or cable and pulls equally on the objects attached at both ends.

For example, in order for a crane to be able to lift the shipping container at constant velocity, the tension in the cable must be at least equal to the weight of the container acting downwards.

Quick Test 4

1. State Newton's first law.
2. State Newton's second law.
3. What is meant by terminal velocity?

Components of forces

Resolving forces into components

Force is a vector quantity, so it has both magnitude **and** direction. Like all vector quantities, two or more forces can be added in a vector diagram (tip-to-tail) to give a resultant force.

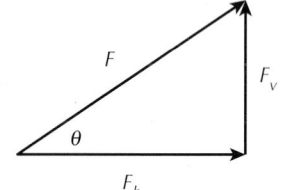

A force acting at an angle can also be split into its perpendicular, horizontal, and vertical components. Trigonometry can be used to work out the magnitude of these components:

$$F_h = F\cos\theta$$

$$F_v = F\sin\theta$$

Example

A box is at rest on a horizontal frictionless surface. A force of 5.0 N is applied at an angle of 25° to the horizontal.

Calculate the horizontal component of this force.

$F_h = F\cos\theta$

$F_h = 5\cos25°$

$F_h = 4{\cdot}5\ N$

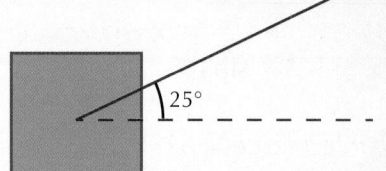

The inclined plane

Inclined plane examples apply to any object placed on a slope.

W is the weight of the object. R is the reaction force which acts perpendicular (normal) to the slope. The force acting parallel to the slope is the resultant of these two forces. This is called component of weight. In the absence of other forces, the component of weight acting parallel to the slope causes the object to accelerate down the slope. It can be shown by trigonometry:

$$parallel\ component\ of\ weight = mg\sin\theta$$

$$normal\ component\ of\ weight = mg\cos\theta$$

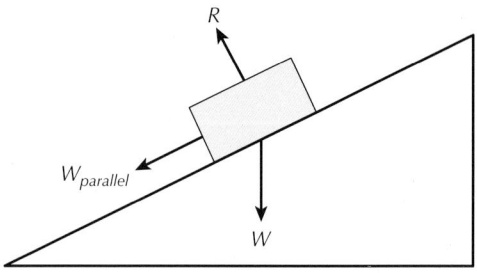

TOP TIP

The equation for the parallel component of weight is not given in the exam relationships sheet.

The inclined plane and friction

In practice there will be some friction between the object and the slope.

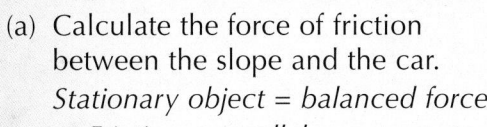

Example

A car mechanic puts a car onto a ramp as shown.

The angle of the slope is increased until the car just remains at rest.

(a) Calculate the force of friction between the slope and the car.
Stationary object = balanced forces
∴ *Friction = parallel component of weight*
parallel component of weight = mgsinθ
= 1000 × 9·8 × sin30°
= 4900 N

(b) The car is 6 m up the ramp. The angle is increased to 40°. Calculate the speed of the car as it slides to the bottom of the ramp.
parallel component of weight = mgsinθ
= 1000 × 9·8 × sin40°
= 6299 N
unbalanced force = 6299 − 4900 = 1399 N
$F = ma$
1399 = 1000a
$a = 1·399\ m\ s^{-2}$
$v^2 = u^2 + 2as$
$v^2 = 0 + 2 × 1·399 × 6$
$v^2 = 16·788$
$v = 4·1\ m\ s^{-1}$

Top Tip

Although the parallel component of weight force will always act down the slope, the force of friction can change its direction. This depends on whether the object is moving up or down the slope. If the block is moving up the slope, then friction will be acting down the slope in the same direction as the component of weight.

Quick Test 5

1. Write down the equations used to calculate the horizontal and vertical components of force.

2. What do we call the force acting parallel to a slope that causes an object to accelerate down the slope?

3. How do we calculate the component of weight?

Energy and power

Work done, gravitational potential and kinetic energy

Work done, gravitational potential and kinetic are all forms of energy measured in joules (J).

Work is done whenever a force is applied to an object to make it move. The quantity of work done by a force can be calculated using the following equation:

> $E_w = Fd$
>
> E_w is the work done in joules (J)
>
> F is the force in Newtons (N)
>
> d is the distance over which the force is applied in metres (m)

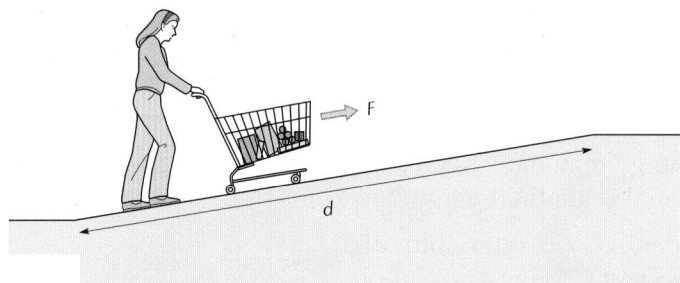

When you lift an object from the ground you have to do work against gravity to raise it to a certain height. In lifting the object, it gains gravitational potential energy. The equation for gravitational potential energy is:

> $E_p = mgh$
>
> E_p is the gravitational potential energy in joules (J)
>
> m is the mass in kilograms (kg)
>
> g is the gravitational field strength in Newtons per kilogram (N kg^{-1})
>
> h is the height in metres (m)

Any object that is moving has kinetic energy. The amount of kinetic energy is determined by the mass of the object and its velocity:

> $E_k = \dfrac{1}{2}mv^2$
>
> E_k is the kinetic energy in joules (J)
>
> m is the mass in kilograms (kg)
>
> v is the velocity in metres per second (m s^{-1})

Power

Power is the rate at which energy is transformed from one form into another.

$$P = \frac{E}{t}$$

P is the power in watts (W)

E is the energy transferred in joules (J)

t is the time taken in seconds (s)

TOP TIP

The units of power can also be expressed as joules per second (J s^{-1}).

Conservation of energy

Conservation of energy states that:

Energy cannot be created or destroyed, only transformed from one form into another.

This means that the total energy at the start of a process will equal the total energy at the end, although the energy may have been converted to other forms, e.g., a falling object converts its gravitational potential energy into kinetic energy.

All forms of energy can be converted into any other form, so it is possible in the majority of cases to equate one energy equation to another.

e.g., $E_p = E_k$

$$\therefore mgh = \frac{1}{2}mv^2$$

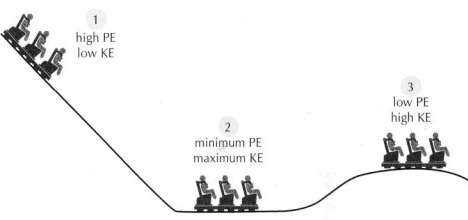

At position 1 the rollercoaster car has a large amount of gravitational potential energy and a low amount of kinetic energy. As the coaster car loses height it gains speed and gravitational potential energy is transformed into kinetic energy. At position 2, the coaster car has reached its maximum kinetic energy and therefore its minimum value of gravitational potential energy. As the coaster gains height it loses speed and kinetic energy is transformed into gravitational potential energy. At position 3, there is a low value of gravitational potential energy and a high value of kinetic energy.

However, at all three positions, the sum of the kinetic and potential energies are equal due to conservation of energy: $(PE_1 + KE_1) = (PE_2 + KE_2) = (PE_3 + KE_3)$.

Quick Test 6

1. What is the unit of energy?
2. State the law of conservation of energy.
3. Explain why an object gains gravitational potential energy when being lifted to a certain height.
4. A box of mass 70 kg is pulled along a horizontal surface by a horizontal force of 90 N. The box is pulled a distance of 12 m. There is a frictional force of 80 N between the box and the surface. Calculate the speed of the box.

Momentum

Where does momentum occur?

Momentum is the product of mass and velocity:

$$p = mv$$

The units of momentum are kg m s^{-1}.

An object can have a large momentum if it is travelling at a large velocity, and also if it has a large mass.

There are many examples in real life that demonstrate momentum, some of which include snooker, curling, rockets and the mechanism of a gun.

As velocity is a vector, so is momentum, and therefore its direction must be taken into account. The momentum has the same direction as the velocity. A change in direction is denoted by a change in sign, i.e., when there is motion in the opposite direction, one momentum must be taken as negative.

In Higher Physics, we need to be able to perform calculations that involve moving objects, as well as objects that diverge into two or more parts. A rocket dropping its boosters is an example of this.

Momentum problems

Calculating the momentum of a single object can be done using the following straightforward equation.

Example

An ice skater has a mass of 65 kg. The ice skater is travelling at 5 m s^{-1} to the right. Find the momentum of the ice skater.

$p = mv$

$p = 65 \times 5$

$p = 325$ kg m s^{-1} to the right

When dealing with collisions or explosions that involve more than one object, it is necessary to calculate the momentum of each individual object.

The following procedure is recommended for all problems involving collisions and explosions:

1. Draw a series of boxes to show the objects before and after the collision or explosion.
2. On each box mark what is known: its mass and/or velocity.
3. If the velocity is known, draw an arrow to indicate its direction.
4. If there is motion in opposite directions, decide which direction is positive and which direction is negative.

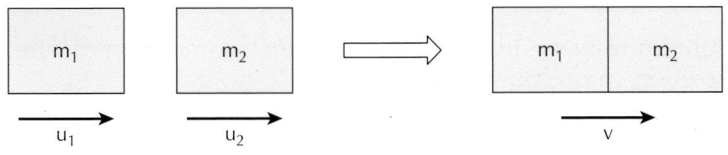

Quick Test 7

1. What is the equation for calculating momentum?
2. What does the momentum of an object depend on?
3. How do we indicate momentum acting in opposite directions?

Conservation of momentum

The law of conservation of linear momentum

The quantity of mass × velocity is the same before a collision and after a collision.

The complete statement of the law of conservation of linear momentum is:

> **The total momentum before a collision is equal to the total momentum after a collision, _in the absence of external forces_.**

It is the total momentum that is conserved; not just the momentum of one object involved in the collision but the sum of all momentums.

Calculations using conservation of momentum

Momentum is conserved in the absence of external forces, and this means that we can calculate unknown masses and velocities.

To do this, we calculate the momentum of each individual object, and equate the total momentum before the collision to the total momentum after the collision. To distinguish between the masses involved, we call them m_1 and m_2. We take a similar approach with the velocities; before the collision we call them u_1 and u_2, and after the collision we call them v_1 and v_2.

Example

A car of mass 400 kg travelling at 5 m s^{-1} collides and locks onto a stationary car of mass 600 kg. Calculate the velocity at which both cars move off.

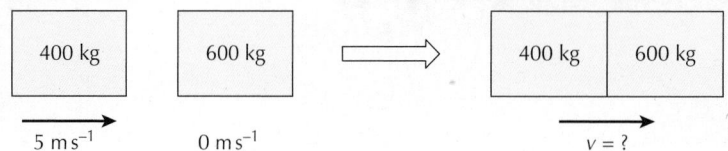

| 400 kg | 600 kg | | 400 kg | 600 kg |

5 m s^{-1} 0 m s^{-1} $v = ?$

Total momentum before = Total momentum after

$$m_1 u_1 + m_2 u_2 = m_1 v_1 + m_2 v_2$$
$$(400 \times 5) + (600 \times 0) = 400v_1 + 600v_2$$
$$2000 + 0 = (400 + 600)v$$
$$2000 = 1000v$$
$$v = 2 \ m \ s^{-1} \text{ to the right}$$

Note: $v_1 = v_2$ so we can use a common velocity of 'v' in the equation. Also, because all motion is in the same direction, there is no need to assign a direction as positive or negative.

TOP TIP

When performing a conservation of momentum calculation involving explosions, the total momentum before is always zero.

Example

A rocket is launched vertically. When it reaches its maximum height it explodes into two parts. Each part is 1500 kg. One part of the rocket moves horizontally at a bearing of 090. Calculate the velocity of the second part of the rocket.

1500 kg	1500 kg		1500 kg	1500 kg

$0 \ m\,s^{-1}$ v_1 $1000 \ m\,s^{-1}$

Total momentum before = Total momentum after

$$0 = 1500 \times v_1 + 1500 \times 1000$$

$$1500v_1 = -1500000$$

$$v_1 = -1000 \ m \ s^{-1}$$

The negative sign indicates that the direction of v_1 is opposite to that of v_2.

Therefore, the second part of the rocket is travelling at 1000 m s^{-1} at a bearing of 270.

Kinetic energy

Momentum is always conserved in a collision, but kinetic energy is not always conserved.

Remember $E_k = \dfrac{1}{2}mv^2$

If momentum and kinetic energy are conserved the collision is said to be **elastic**. In this case:

$$E_k \ before = E_k \ after$$

If momentum is conserved but kinetic energy is not conserved the collision is said to be **inelastic**. In this type of collision:

$$E_k \ before \neq E_k \ after$$

Kinetic energy can be 'lost' during a collision to other forms of energy, such as sound and heat.

Calculations to determine whether a collision is elastic or inelastic can be carried out in a similar way to conservation of momentum calculations. The difference lies in the calculation of the kinetic energy of each individual object, which is then compared to the total kinetic energy both before and after the collision. If the values of kinetic energy are equal the collision is elastic. If the values of kinetic energy are not equal, it can indicate either an inelastic collision or an explosion.

$$E_k \ before = \frac{1}{2}m_1u_1^2 + \frac{1}{2}m_2u_2^2$$

$$E_k \ after = \frac{1}{2}m_1v_1^2 + \frac{1}{2}m_2v_2^2$$

Quick Test 8

1. State the law of conservation of linear momentum.
2. Describe the differences between elastic and inelastic collisions.

Impulse

Change in momentum

From Newton's second law:

$$F = ma = \frac{m(v-u)}{t} = \frac{mv - mu}{t}$$

This expression states that the unbalanced force acting is equal to the rate of change of momentum.

Impulse

Impulse is the product of force and time, and is the cause of change in momentum. Impulse is measured in Newton seconds, N s. Rearranging the equation from Newton's second law gives:

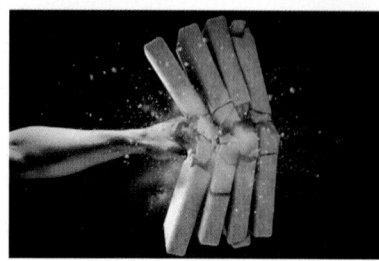

$$Ft = mv - mu$$

where F is the unbalanced force in Newtons

t is the time of impact in seconds

u is the initial velocity before impact in metres per second

v is the final velocity after impact in metres per second

Impulse is equal to the change in momentum. Thus, the change in momentum depends on the force of impact, and the time over which the force acts.

Impulse is a vector quantity, and so care must be taken in the correct use of positive and negative signs for the values of u and v.

TOP TIP

Change in momentum can also be written as Δp.

Force–time graphs

It is possible to graph force against time for an impact. Ideally:

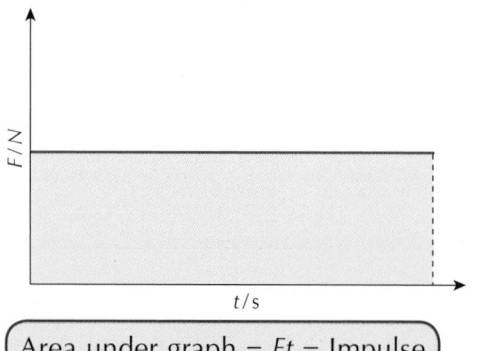

Area under graph = Ft = Impulse

In reality, the force applied is not usually constant:

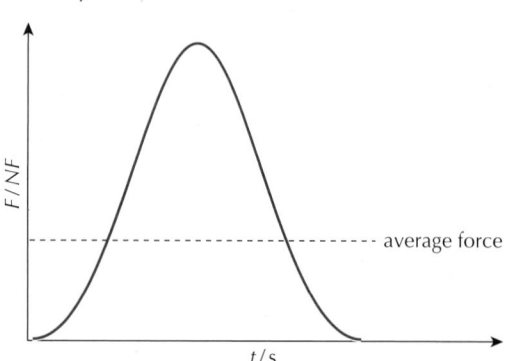

average force

Changing the force of an impact

For the same change in momentum, if t is made bigger then F gets smaller. The opposite is also true for the same change in momentum: if t is made smaller then F gets bigger.

Modern automotive engineering is a good example of this idea in practical terms. Cars are designed with crumple zones on their front and rear. These specially designed areas absorb the energy of a collision, and increase the amount of time it takes for the car to come to a stop. For a given change in momentum, the time of impact t increases. The effect of this is that the force of impact F acting on the passengers is reduced. This increases the chance of surviving a serious incident.

Similarly, for a given change in momentum, a softer surface will apply a smaller force than a harder surface. The softer surface can apply a smaller average force over a much longer period of time, as the material deforms more readily than the harder surface. This results in the same change in momentum, but with a much less damaging force.

Momentum and Newton's third law

It can be shown that conservation of momentum and Newton's third law are equivalent.

Start with conservation of momentum:

$$total\ momentum\ before = total\ momentum\ after$$

$$m_1u_1 + m_2u_2 = m_1v_1 + m_2v_2$$

$$m_2u_2 - m_2v_2 = m_1v_1 - m_1u_1$$

$$-m_2(v_2 - u_2) = m_1(v_1 - u_1)$$

i.e., − (change in momentum of object 2) = (change in momentum of object 1)

Apply the impulse relationship:

$$-F_2t = F_1t$$

$$so\ -F_2 = F_1$$

This proves the link between Newton's third law and momentum. Newton's third law states that if object 1 exerts a force on object 2, then object 2 exerts an equal force on object 1 that is opposite in direction.

Quick Test 9

1. What dictates the rate of change of momentum of an object?
2. How can you calculate the impulse from a force–time graph?
3. Name two ways to reduce the force of impact for a given change in momentum.
4. State Newton's third law.

Horizontally launched projectiles

Projectiles

A projectile is an object which is influenced only by the downward force of gravity. Once projected, it continues its motion under its own inertia. Projectiles can be launched horizontally, vertically or a combination of the two.

Examples of projectiles are objects thrown vertically upwards; objects dropped vertically downwards; objects launched upwards at an angle or objects launched from a height horizontally.

> **TOP TIP**
>
> In projectile motion, ignore all air resistance and any other forces, except gravity.

Horizontal and vertical components

Projectiles have both horizontal and vertical components of motion. As the only force acting on the projectile is gravity, which acts in one direction, only one of the components is being acted upon by a force. This means that the two components are moving differently, and therefore must be treated separately.

Projectiles fired horizontally

An object dropped vertically and one thrown horizontally will fall at the same rate. This is because they are both subject to the same force of gravity in the vertical direction.

If we look at the two different components of the motion:

Horizontally: there are no forces acting so the horizontal velocity is constant.

Vertically: the object experiences a constant acceleration due to the force of gravity.

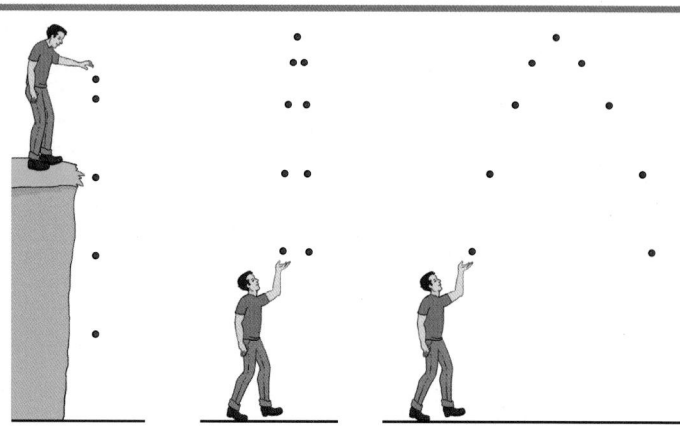

gravity-free path

projectile motion

vertical motion only

The combination of these two motions produces the curved path of the projectile.

Example

75 m s⁻¹

A plane is travelling with a constant horizontal velocity of 75 m s⁻¹ when a box is dropped out of it. The box lands on the ground after a time of 15·5 seconds.

(a) What is the horizontal distance travelled by the box after it is released?

(b) What is the vertical velocity of the box when it hits the ground?

Solution:

Horizontal	Vertical
s = ?	s = ?
v = 75 m s⁻¹	u = 0 m s⁻¹
t = 15·5 s	v = ?
	a = 9·8 m s⁻²
	t = 15·5 s

(a) $s = \bar{v}t$
$s = 75 \times 15 \cdot 5$
$s = 1163 \, m$

(b) $v = u + at$
$v = 0 + 9 \cdot 8 \times 15 \cdot 5$
$v = 151 \cdot 9 \, m \, s^{-1}$

> **TOP TIP**
>
> The initial vertical velocity of a horizontal projectile is 0 m s⁻¹.

Quick Test 10

1. Describe what is meant by a projectile.
2. Name the two components of a projectile's motion.
3. Is the following statement true or false:
   Two identical objects are launched from the same height at the same time. One falls vertically and the other horizontally when released. The vertical object strikes the ground first.
4. Describe the horizontal motion of a projectile.
5. Describe the vertical motion of a projectile.

Oblique projectiles

Projectiles at an angle

Projectiles launched at an angle are called oblique projectiles. As with horizontal projectiles, they are only acted upon by the force of gravity.

The horizontal and vertical components must be treated separately. The velocity at an angle must be split into its vertical and horizontal components before any calculations can be attempted.

Horizontal and vertical components

Any vector can be resolved into its horizontal and vertical components. For an oblique projectile:

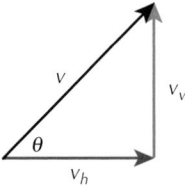

where v is the velocity at an angle θ, v_h is the horizontal component of velocity, and v_v is the vertical component of velocity.

Trigonometry can be used to find the values of the horizontal and vertical components. For the horizontal component:

$$\cos\theta = \frac{adjacent}{hypotenuse}$$

$$\cos\theta = \frac{v_h}{v}$$

$$\therefore \quad v_h = v\cos\theta$$

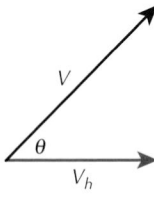

For the vertical component:

$$\sin\theta = \frac{opposite}{hypotenuse}$$

$$\sin\theta = \frac{v_v}{v}$$

$$\therefore \quad v_v = v\sin\theta$$

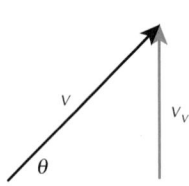

Both of these components combine to produce the curved path of the projectile.

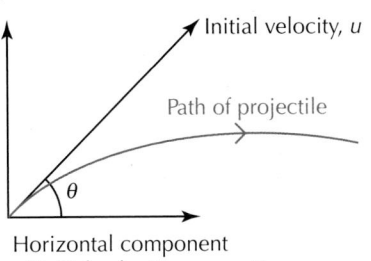

Vertical component of initial velocity = $u \sin \theta$

Path of projectile

Initial velocity, u

Horizontal component of initial velocity = $u \cos \theta$

Example

A projectile is fired at 40 m s^{-1} at an angle of 30° to the horizontal.

Calculate:

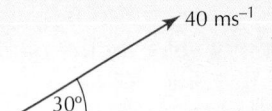

(a) The horizontal component of the velocity

$v_h = v\cos\theta$
$v_h = 40\cos30°$
$v_h = 34{\cdot}6 \ m \ s^{-1}$

(b) The vertical component of the velocity

$v_v = v\sin\theta$
$v_h = 40\sin30°$
$v_h = 20 \ m \ s^{-1}$

Calculations involving oblique projectiles

For projectiles launched at an angle:

1. The vertical velocity at the maximum height is zero.
2. The initial vertical velocity is $v \sin\theta$.
3. The horizontal velocity is $v \cos\theta$ and is constant throughout.
4. If take-off and landing heights are the same then:
   (a) The time to rise is the same as the time to fall.
   (b) The initial vertical velocity is equal to the final vertical velocity, but in the opposite direction, i.e., $u_v = -v_v$.

Quick Test 11

1. How do you calculate the horizontal component of velocity for an oblique projectile?
2. How do you calculate the vertical component of velocity for an oblique projectile?
3. What is the vertical velocity of an oblique projectile when it reaches its highest point?
4. What can be said about the time the projectile takes to rise and the time it takes the projectile to fall, if it falls back to the same level?

Satellite motion

Newton's thought experiment

Isaac Newton conducted a thought experiment where he imagined a cannon ball being fired horizontally off a cliff. The cannon ball follows a curved path as the force of gravity acts to make it fall back to Earth. If the projectile is launched with a greater horizontal velocity it lands further from the cannon. If the ground also curves away from the cannon it will land even further from where it was launched.

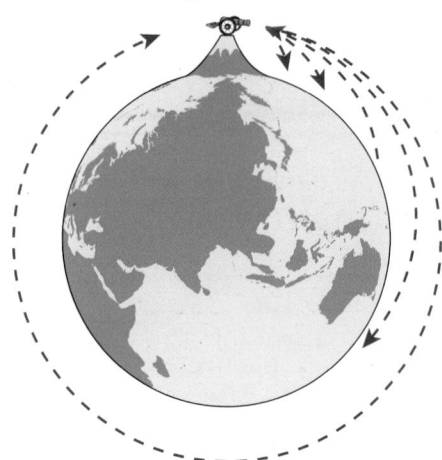

Newton imagined that there was a horizontal launch velocity that would result in the projectile falling to Earth at the same rate at which the Earth's surface curves away from it. It would therefore orbit the Earth. At this velocity the cannonball would never return to Earth, unless acted upon by an external force.

Newton's thought experiment suggested satellite motion roughly 300 years before Sputnik 1, the first man-made object to orbit the Earth.

Satellites in orbit

The speed of a satellite in orbit around the earth is constant. However, the satellite obeys Newton's second law of motion: it is in orbit, meaning its direction is constantly changing as gravity pulls the satellite towards the earth.

Remember that velocity is a vector quantity, requiring both speed and direction, which means the satellite's velocity is constantly changing and it is therefore accelerating.

The satellite is said to be in free-fall. An object is described as being in free fall when the force of gravity is the only force acting on it.

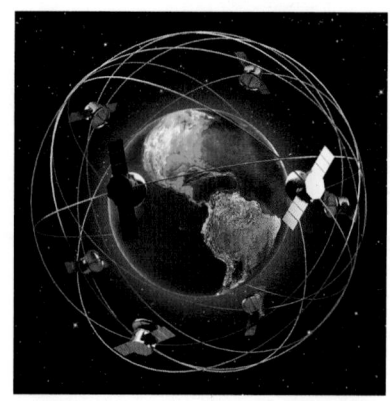

The speed of the satellite needs to be just right to remain in orbit. If the satellite is too slow, it will fall to earth. If it is too fast, it will leave Earth's orbit completely.

TOP TIP

In the absence of friction both projectiles and objects in freefall are acted upon by only one force, the force of gravity.

Types of satellites

Different types of satellites orbit the Earth at different radii and therefore have different orbital periods. This also means that they have different orbital velocities.

Satellites in Low Earth Orbit (LEO), such as the International Space Station, orbit at a radius of up to 2000 km, with a period of approximately 90 minutes.

Many communication satellites, including TV satellites, orbit at a radius of 36 000 km and have a period of 24 hours. These are called geostationary satellites, and stay above the same point on the Earth's surface. Because these satellites stay in a fixed position relative to the Earth, domestic satellite dishes do not have to 'track' these satellites across the sky.

Quick Test 12

1. What quantity determines whether a satellite stays in orbit around the Earth?

2. State what is meant by free fall.

3. List two types of orbit that a satellite can have.

Newton's universal law of gravitation

The law of universal gravitation

Newton's law of gravitation states that the gravitational attraction that exists between two objects is directly proportional to the mass of each object, and inversely proportional to the square of their distance apart.

$$F = \frac{Gm_1m_2}{r^2}$$

G is the universal gravitational constant and has the value $6\cdot67 \times 10^{-11} \text{m}^3 \text{ kg}^{-1} \text{ s}^{-2}$

m_1 and m_2 are the two masses, measured in kilograms

The distance, r, between the two objects is the distance between their **centres of mass**. A common mistake is to measure from the surfaces of the objects (planets, moons, etc.).

TOP TIP

The value for G can be found in the data sheet at the front of your exam paper.

Inverse square law

Beyond the surface of an object, the gravitational force decreases with increasing distance. The gravitational force varies inversely with the square of the distance, i.e., doubling the distance decreases the force by a factor of 4. This is one of many inverse square laws in physics.

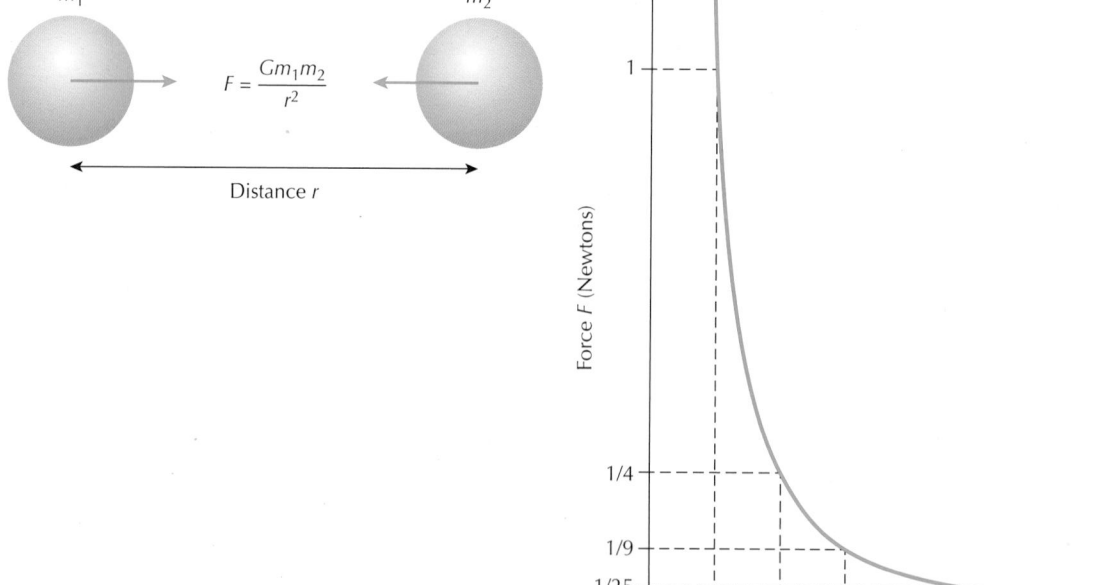

The force of gravity in everyday life

Newton's law of gravitation can be used to calculate the value of the force between any two objects of any size. Everyday objects exert a force of gravitational attraction on one another; the reason we do not notice these forces in daily life is because these gravitational forces are so small. The effects of these forces are only apparent when very large masses are involved.

Gravitational field strength

Every mass causes a gravitational field around itself. A gravitational field is a region where other masses will experience a gravitational force. The gravitational field strength at a point is the force per unit mass exerted on a mass placed at that point. This means that the gravitational field strength is equal to the force experienced by a mass of 1 kg in that gravitational field.

As g is the force per unit mass, it is related to the mass, M, of the object providing the force by the following expression:

$$g = \frac{F}{m} = \frac{GM\cancel{m}}{\cancel{m}r^2} = \frac{GM}{r^2}$$

Like gravitational force, beyond the surface of the object, the value of g follows an inverse square law.

Gravitation and the universe

The application of Newton's law of gravity has enabled us to acquire much of the detailed information we have about the planets in our solar system, the mass of the Sun, and even the existence of dark matter. We know the masses of planets and stars that far outreach our current travelling abilities; these masses are obtained by applying the laws of gravity to the measured characteristics of the orbit. In space an object maintains its orbit because of the force of gravity acting upon it. Planets orbit stars, stars orbit galactic centres, galaxies orbit a center of mass in clusters, and clusters orbit in superclusters. All of this is due to the universal law of gravitation, which dictates the orbit of all the planets in our solar system, and beyond.

Quick Test 13

1. List the two variables that determine the gravitational force of attraction between two objects.

2. Why do we not notice gravitational forces in everyday life?

3. State what is meant by gravitational field strength.

4. Give two applications of the universal law of gravitation in the universe.

Galilean invariance and Newtonian relativity

Galilean invariance

Galilean relativity or Galilean invariance states that the laws of motion remain the same whether you are moving steadily, or at rest.

Galileo Galilei used a ship moving at constant velocity, without rocking, on a smooth sea to conclude that any observer conducting experiments below the deck would be unable to determine whether the ship was moving or stationary. In other words, when you are moving at a constant speed, the laws of physics are exactly the same as when you are stationary.

Newtonian relativity and absolute space

Newtonian physics assumes the concept of absolute space and time.

Absolute space implies a static backdrop against which all movement can be referenced. For example, think about sitting on a train next to another train when one of the trains starts moving. It is sometimes difficult to tell which of the two trains is moving, until you relate the movement to the fixed background of the train station. Newton assumed that there was always a fixed reference point for all moving objects. This fixed reference point is known as absolute space.

Absolute time assumes that time is the same for all observers, including those in motion. This absolute time exists across the universe, meaning that an observer on Mars would experience time in the same way as an observer on Earth.

In a Newtonian universe, there should be no difference in space or time regardless of where you are, or how fast you are moving. Therefore, two observers should agree on measurements of distance and time.

TOP TIP

The moving observer must be travelling at a constant speed in order to agree with the stationary observer on measurements of distance and time.

Reference frames

A frame of reference is a physics term that describes the measurements made by a particular observer. Relativity is chiefly concerned with making observations and measurements from different reference frames.

Imagine three different observers: a passenger sitting next to you on a train travelling east at 60 mph, a passenger on another train travelling west at 40 mph and a person standing stationary on a train platform. Each observer is in their own frame of reference so will make their own observations about your motion.

Observer	Location	Observation
1	Person standing on the train platform	You are travelling towards them at 60 mph
2	Passenger on a train travelling west at 40 mph	You are travelling towards them at 100 mph
3	Passenger sitting next to you	You are stationary

Quick Test 14

1. What is meant by Galilean invariance?

2. True or false? In a Newtonian universe, time is different for moving observers than for stationary observers.

3. Imagine you are stationary on a train platform. A train approaches the platform at 20 m s^{-1}. An observer is walking towards the front of the train at a constant speed of 2 m s^{-1}.

   a) What observation do you make?

   b) What observation does the observer make?

Special relativity

The principles of relativity

Special relativity is described as 'special' because it only considers the 'special' case where reference frames are moving at a constant speed.

Einstein built his special theory of relativity by using two basic ideas (or postulates):

- The laws of physics are the same for all observers.
- The speed of light (in a vacuum) is the same for all observers.

This means that light does not obey the ordinary laws of relative motion. The speed of light does not depend on the motion of the source, or the motion of the observer towards or away from the source.

In order for the speed of light to remain constant, there must be changes in the measurements made of distance and of time.

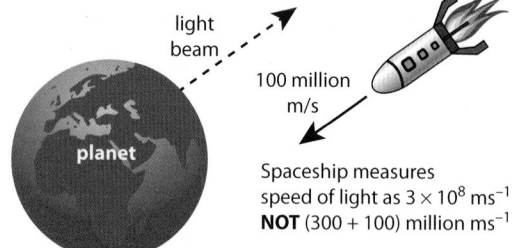

light beam

100 million m/s

planet

Spaceship measures speed of light as 3×10^8 ms^{-1}
NOT (300 + 100) million ms^{-1}

Time dilation

One consequence of the speed of light being the same for all observers is that the time experienced by all observers is not always the same. A process that takes a certain time to occur in a moving frame of reference is observed to take a **longer** time from the point of view of someone in a different frame of reference.

For example, the stationary observer outside the moving spacecraft can see a clock on the moving spacecraft move its second hand 10 seconds, but can also see that this takes 11 seconds when measured by her own watch. The observer in the moving spacecraft also sees their own watch has counted 11 seconds, but measures the observer outside the spacecraft as having a time of 10 seconds on their watch. Both observers conclude that the other person's watch is running slowly.

11 s

10 s

10 s

11 s

Time dilation and the Lorentz factor

$$t' = \frac{t}{\sqrt{1-\left(\frac{v^2}{c^2}\right)}}$$

t = time interval in a frame of reference
t' = time interval measured by an observer in a **different** frame of reference
v = the relative velocity of the two frames of reference
c = the speed of light in a vacuum

This equation is often written as:

$$t' = t\gamma$$

where γ is the Lorentz factor

$$\gamma = \frac{1}{\sqrt{1 - \left(\frac{v^2}{c^2}\right)}}$$

A graph of the Lorentz factor against speed shows the dramatic effects of relativity as v approaches the speed of light.

This is the reason we do not experience time dilation in daily life, as the Lorentz factor is only significant for extremely fast speeds.

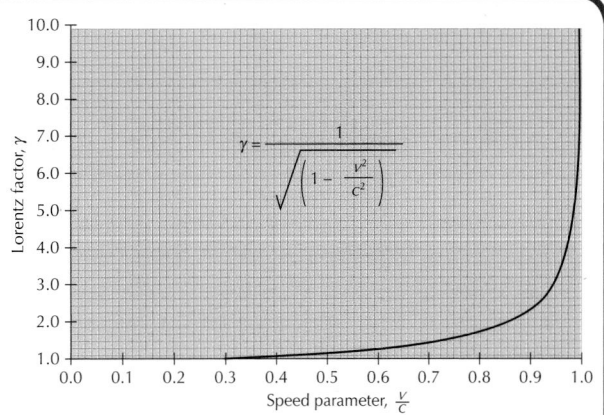

TOP TIP

The speed of light in a vacuum is 3×10^8 m s^{-1}.

Length contraction

Another consequence of the speed of light being the same for all observers is the shortening or contraction of length when an object is moving. An object moving at an extremely high speed will appear shorter than an identical object which is stationary, when both are observed by the same stationary observer. For example, the length of a rocket travelling at 1×10^8 m s^{-1} will appear shorter in length when compared to an identical rocket stationary on Earth, when both are observed by the same stationary observer on Earth.

$$l' = l\sqrt{1 - \left(\frac{v^2}{c^2}\right)}$$

l = distance in a frame of reference

l' = distance measured by an observer in a **different** frame of reference

v = the relative velocity of the two frames of reference

c = the speed of light in a vacuum

Quick Test 15

1. What are the two postulates of special relativity?

2. The speed of light must be the same for all observers. Name two consequences of this.

3. A rocket ship travels at a constant speed of $0.8c$ relative to the Earth. A clock on the rocket ship records a journey time of 4.8 hours. What would the flight time be when measured by a clock on Earth?

Introducing the Doppler effect

The Doppler effect

The Doppler effect is the change in frequency of a wave when the source or the observer is moving. It can be observed in all types of waves, including light and all the waves in the electromagnetic spectrum.

The Doppler effect is most commonly experienced with sound waves. A fast moving car approaching and then receding emits a sound that is of a higher frequency as it approaches, and a lower frequency as it moves away. The same effect can be observed with a passing train or aeroplane.

Sound and the Doppler effect

When a sound source is stationary and emitting sound of a constant frequency, all observers will hear the same frequency. This is because the wavefronts are always the same distance apart, meaning that the wavelength (λ), and therefore also the frequency, remain constant, provided the observers are also stationary.

In this course, we only deal with a wave source moving at a constant speed relative to a stationary observer. If the sound source is moving, the stationary observer will hear changes in the frequency of that sound. What these changes are will depend upon on whether the sound source is moving towards or away from the stationary observer.

When the source moves towards the observer, the observed frequency increases.

When the source moves away from the observer, the observed frequency decreases.

This is because as the sound waves are emitted, the sound source moves some distance to the right. This causes the wavefronts to become more spread out on the left (λ_1) and closer together on the right (λ_2).

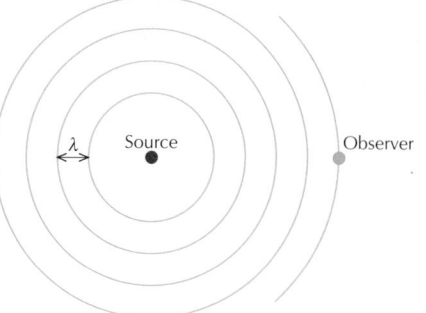

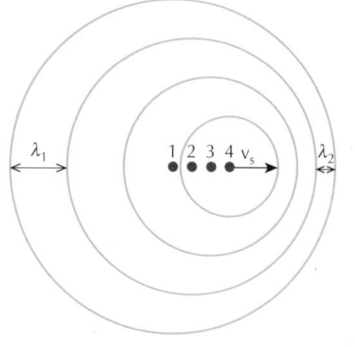

● = position of moving source S

Calculations involving sound

For the source moving **towards** the observer:

$$f_{observed} = f_{source}\left(\frac{v}{v - v_{source}}\right)$$

For the source moving **away** from the observer:

$$f_{observed} = f_{source}\left(\frac{v}{v + v_{source}}\right)$$

The equation given in the relationships sheet is:

$$f_o = f_s\left(\frac{v}{v \pm v_s}\right)$$

It is important that you remember whether to use the '+' or '−' appropriately. Ensure that your answers give the expected increase or decrease in frequency.

TOP TIP

v = speed of the wave, e.g., sound or light.

Example

A police car emits sound waves with a frequency of 1000 Hz from its siren. The car is travelling at 20 m s^{-1}.

Calculate the frequency heard by a stationary observer as the police car moves towards them.

$$f_o = f_s\left(\frac{v}{v - v_s}\right)$$

$$f_o = 1000\left(\frac{340}{340 - 20}\right)$$

$$f_o = 1062 \cdot 5\,Hz$$

Quick Test 16

1. Identify the types of waves that display the Doppler effect.

2. Does the observed frequency increase or decrease as the sound source moves away from the observer?

3. Does the observed frequency increase or decrease as the sound source moves toward the observer?

4. An observer is standing on a station platform. A train approaching the station sounds its horn as it passes through the station. The train is travelling at a speed of 25 m s^{-1}. The horn has a frequency of 200 Hz. Calculate the frequency heard by the observer as the train is moving away from the station.

Redshift

Light from distant galaxies

As the universe is expanding, light from distant galaxies shows the Doppler effect. This causes the light emitted to shift towards the red end of the electromagnetic spectrum, due to an increase in wavelength. This is called **redshift**, and can be used to measure very large distances, as well as the speed at which galaxies are receding (moving away).

If galaxies are moving towards the Earth, this causes the light emitted to shift towards the blue end of the spectrum. This is called **blueshift**, and can again be used to measure distances in space, and the speed at which the galaxies are moving towards us.

Line spectra

Redshift can be observed in the line spectra of light coming from distant stars. Line spectra display how an element absorbs and emits light energy. These are known as absorption and emission lines.

Each element has its own unique absorption and emission lines. This means that the light observed from distant stars and galaxies can be used to identify the elements present in that star or galaxy. Because the absorption and emission lines are constant throughout the universe, the line spectra from distant objects can be compared with the spectra observed here on Earth.

If the lines in a star or galaxy's absorption or emission spectrum are shifted towards the red end of the spectrum (longer wavelength), then the star or galaxy is moving away from the Earth. If the star or galaxy is moving towards the Earth, then the lines in the absorption or emission spectrum are shifted towards the blue end of the spectrum (shorter wavelength).

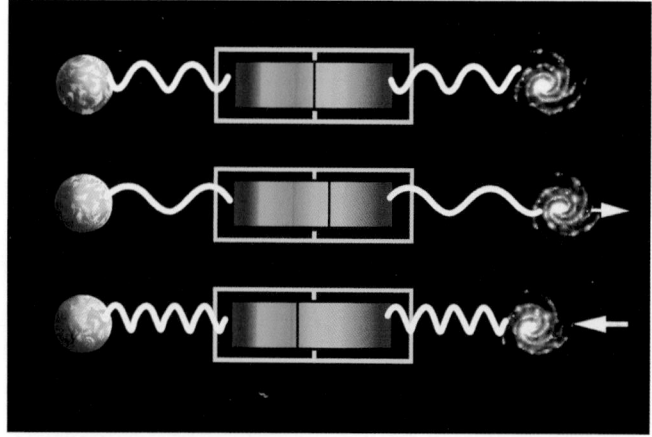

Calculating redshift

Redshift, z, of a galaxy is given by the change in the wavelength of an absorption line divided by the original wavelength:

$$z = \frac{\lambda_{observed} - \lambda_{rest}}{\lambda_{rest}}$$

Redshift of galaxies travelling at non-relativistic speeds (i.e., less than 10% of the speed of light) can also be calculated from the ratio of the velocity of the galaxy to the velocity of light:

$$z = \frac{v_{galaxy}}{c}$$

TOP TIP

Since the redshift, z, is a ratio of two identical quantities it does not require units.

Fast moving galaxies

The Doppler effect equations for sound cannot be used with light from fast moving galaxies. This is because relativistic effects need to be taken into account.

Because the source of the light is moving at very high speeds an observer on Earth would perceive the galaxy to be moving more slowly, as a result of time dilation. This in turn means that the observed frequency of the light will be lower, and is thus shifted towards the red end of the spectrum. Therefore, for a fast moving galaxy the light is redshifted because of the Doppler effect **and** time dilation. Einstein made adjustments to the Doppler effect equations to take time dilation effects into account.

Quick Test 17

1. How does redshift affect the wavelength of light from distant galaxies?
2. What does redshift indicate about the motion of distant galaxies?

Hubble's law

What is Hubble's law?

From his observations of the redshift of galaxies, Edwin Hubble discovered that most of them were moving away from us. He also found that the further away any galaxy was, the faster it was moving. The relationship between distance and the speed of a galaxy is known as Hubble's law.

The graph adjacent shows a plot of the velocities of a number of galaxies (the black dots) as they are observed to move away from Earth. Looking at the plotted points, a trend (the red line) can be derived. The gradient of this line is called 'Hubble's constant'.

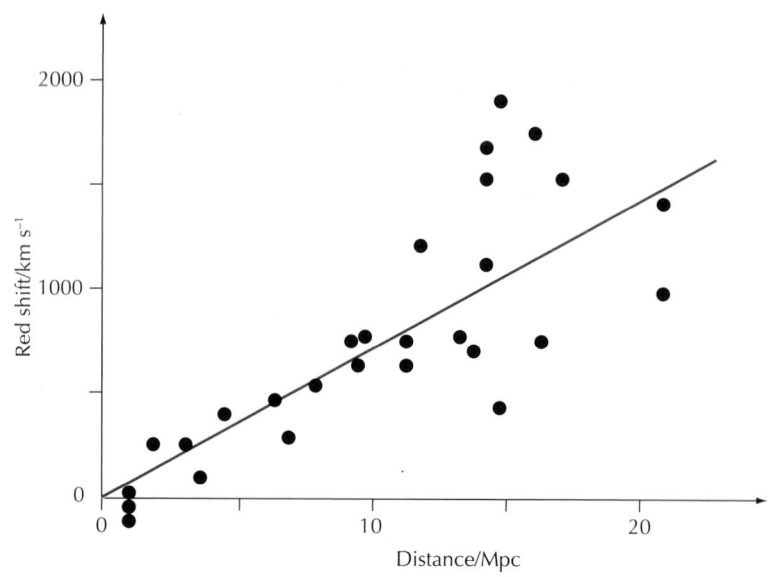

TOP TIP

A Megaparsec (MPC) is a huge distance. It is equivalent to 3.26 light years, which is approximately 30,800,000,000,000,000,000,000 km.

Hubble's constant (H_o)

The current, measured, rate of expansion of the universe is called Hubble's constant (H_o). This is calculated by

$$v = H_o d$$

Where v = the recession velocity of a galaxy (in ms^{-1})

      d = the distance of the galaxy from Earth (in m)

      H_o = Hubble's constant (in s^{-1})

TOP TIP

The value for Hubble's constant can be found in the data sheet at the front of your exam paper.

Its value currently stands at $2 \cdot 3 \times 10^{-18}$ s^{-1}. There has been much scientific debate over the accuracy of this number. As more precise measurements are taken, this figure is likely to be updated, potentially revealing that the 'constant' is, in fact, increasing over time. Hubble's original data, gathered in the 1920s, went to distances of around 2MPC, with corresponding velocities of approximately 1000 kms^{-1}. Modern data takes this distance to over 700 MPC and corresponding velocities of 4×10^4 kms^{-1}. This improvement in the set of data has created a great deal of discussion. If Hubble's constant is measured to be 'ever increasing' it forces scientists to look into why this might be happening. One theory is that the constant is increasing due to Dark Energy (see page 38).

The age of the universe

Hubble's observations showed that the universe is expanding. Assuming that this rate of expansion has remained constant throughout all time, it is possible to estimate the age of the universe, by working back in time to when all the galaxies were at the same point in space.

If in a time, t, a galaxy has moved outwards by a distance, d, at velocity, v, then:

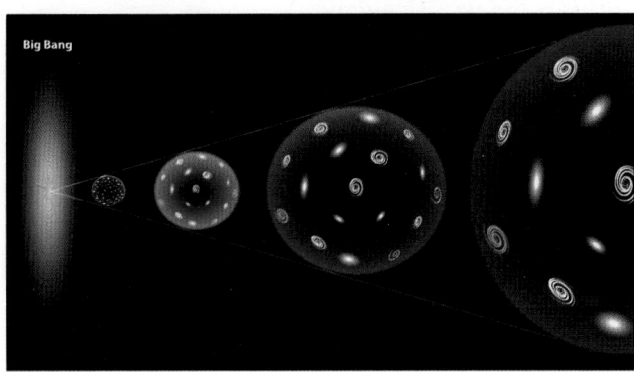

Big Bang

$$t = \frac{d}{v}$$

But from Hubble's law:

$$\bar{v} = H_o d$$

Therefore:

$$t = \frac{d}{H_o d}$$

$$t = \frac{1}{H_o}$$

Current estimates using this method place the age of the universe at approximately 13·8 billion years.

Quick Test 18

1. What is meant by the recession velocity of a galaxy?
2. Hubble's law shows the relationship between which two quantities?
3. Use Hubble's constant to calculate the age of the universe.

Expansion of the universe

Rate of expansion

Edwin Hubble's measurements of the velocity of galaxies suggested that the universe was expanding. It has been shown in recent years that the rate of this expansion is increasing. Observations of distant supernovae have shown that they are less bright than expected, indicating that the rate of expansion is not slowing, but accelerating. Measurements made of cosmic microwave background (see page 42) provide further evidence for an accelerating universe.

Dark energy

The force of gravity holds galaxies together. Gravity should therefore be an unbalanced force that acts to slow the rate of expansion of the universe, yet observations have shown that this is not the case. This suggests that there is a force acting to overcome the gravitational attraction between galaxies. This force must be significantly greater than the force of gravity in order to cause the accelerated expansion of the universe.

Astronomers and cosmologists have as yet been unable to determine the source of energy for this mysterious force, so it is referred to as dark energy.

The magnitude of the force of gravity is determined by the mass in the universe. If the mass of the universe is large enough that the force of gravity is greater than the force produced by dark energy, the universe will begin to collapse in on itself, ending in a 'Big Crunch'. If the mass of the universe is small enough that the force of gravity is less than the force produced by dark energy, the universe will continue expanding forever.

Dark matter

Inside a galaxy, stars moving towards us will show blueshift (shorter wavelength), and stars moving away from us will show redshift (longer wavelength). It is possible to use these measurements to determine the orbital speeds of stars. Stars that are further out in a galaxy should travel slower than those closer to the galactic core, as they experience a smaller gravitational force.

The graph here shows that the velocity of stars remains almost constant as the distance from the centre of the galaxy increases. Stars in these galaxies moving at these speeds cannot be held together by normal matter. There must therefore be a large amount of mass that cannot be seen. This 'missing mass' is called dark matter.

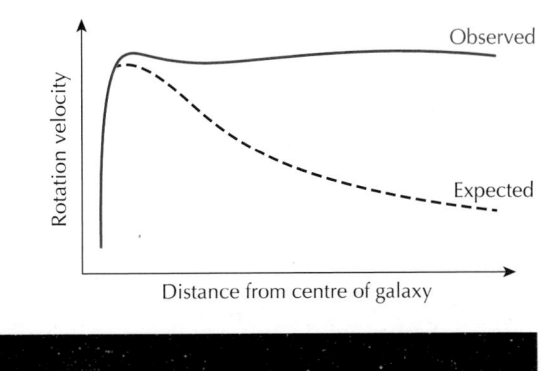

TOP TIP

Dark matter emits no radiation, and so is extremely difficult to measure.

Quick Test 19

1. What is happening to the rate of expansion of the universe?
2. What is causing the universe to continue expanding?
3. What is the 'missing mass' in galaxies referred to as?

Blackbody radiation

The temperature of stellar objects

We know from various everyday experienes that objects glow when heated to high temperatures.

As an object becomes hotter it glows red, then orange, then yellow, then white. Finally, when the temperature is extremely high it glows a blue-white colour. If you look closely at stars in the night sky you will see that they also have a colour. The colour of a star allows us to determine the surface temperature of the star. The hottest stars appear blue-white with the lower temperature stars appearing red in colour.

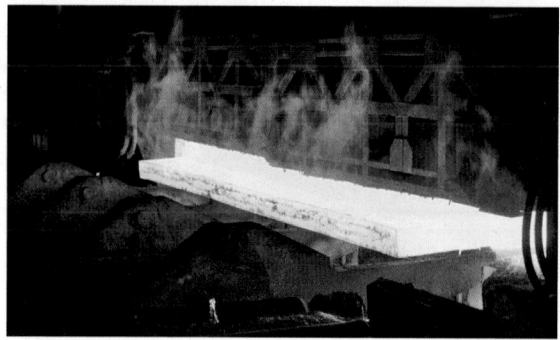

TOP TIP

Don't confuse the order of the wavelengths of visible light (ROYGBIV) with star colours.

Blackbody

An object that absorbs all the electromagnetic radiation that falls on it is also the **best** emitter of electromagnetic radiation of any wavelength. Such an object is called a blackbody, and it is a perfect emitter and absorber of radiation. The continuous spectrum of radiation that it emits is called blackbody radiation.

Hot objects such as stars behave effectively like a blackbody and emit blackbody radiation. The distribution of energy is spread across a wide range of wavelengths; however, the peak wavelength (colour) gives us the surface temperature of the star.

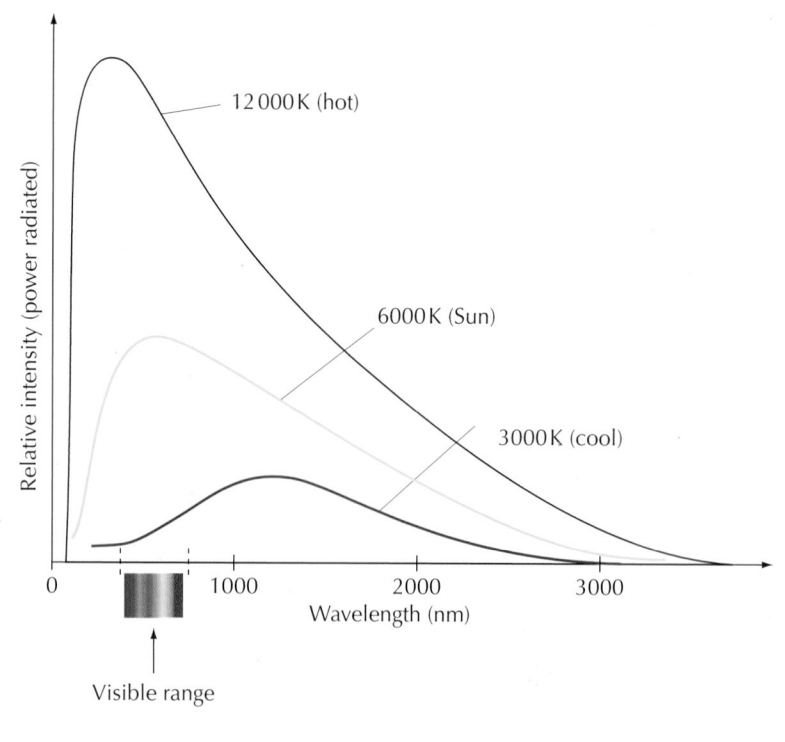

When the temperature is increased, the peak moves towards the short wavelength end. This gives the change in colour from red to orange to yellow to white to blue-white.

Hotter stars emit more radiation per unit surface area. Remember that the radiation emitted is not only made up of visible light, but occurs over the **full range** of the electromagnetic spectrum from radio waves to gamma rays. Hotter stars also emit higher energy photons, meaning they have a shorter wavelength and higher frequency. This relates to the increase in energy with frequency in the electromagnetic spectrum.

The wavelength of the peak of the blackbody radiation curve decreases with increasing temperature. The fourth root of the intensity shows clearly the variation in wavelength.

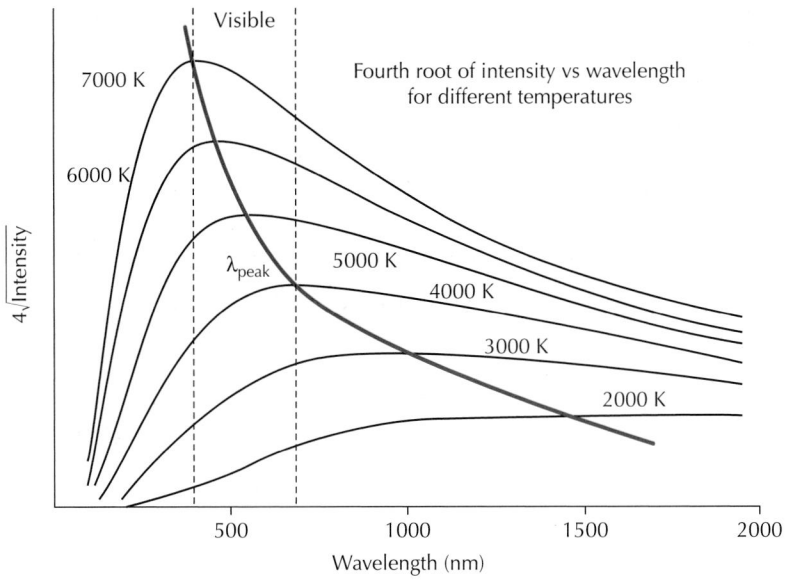

Quick Test 20

1. List the colours of stars from the lowest to the highest temperature.
2. How is the surface temperature of a star determined?

Evidence for the Big Bang

Big Bang theory

If we picture the expanding universe in reverse, it isn't hard to imagine that the universe began from a singularity. In the beginning, the universe was very small and incredibly hot. As the universe expanded it began to cool down and eventually atoms of hydrogen and helium were able to form. These collections of atoms later went on to form stars and galaxies.

The cosmic microwave background

If the Big Bang did happen, there should be some radiation left over that can be detected in the present day. The electromagnetic radiation produced at the Big Bang would be redshifted due to the expansion of the universe. It was predicted that the peak wavelength associated with this radiation should be in the microwave region and correspond to a temperature of around 2·7 kelvin. The radiation should be the same in all directions and was therefore called the cosmic microwave background.

It was not until the 1960s that two astronomers, Penzias and Wilson, accidentally detected the cosmic microwave background. Today, highly accurate measurements from the Cosmic Background Explorer (COBE) satellite have allowed us to measure the cosmic microwave background in more detail, and confirm a peak wavelength that corresponds to a temperature of around 2·7 kelvin.

This measurement of 2·7 kelvin is consistent with a universe that expanded and cooled over a period of around 13·8 billion years.

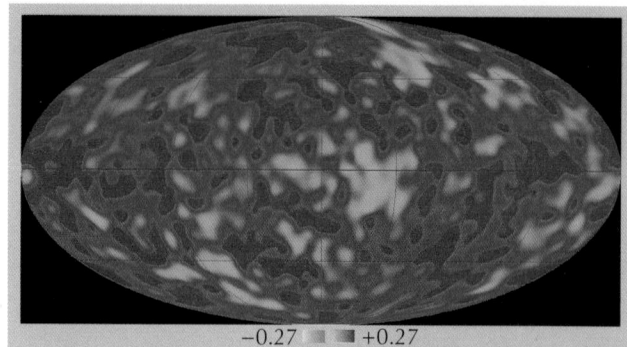

Abundance of lighter elements

The most abundant elements in the universe are hydrogen (70–75%) and helium (25–30%). This amount of hydrogen and helium cannot be explained purely by nuclear fusion processes in stars. However, it can be accounted for by the formation of lighter elements shortly after the Big Bang.

Olber's paradox

When we view the night sky on a clear night we see huge numbers of stars, but we also see areas of darkness between them. Given that there are countless stars in the universe, all similar in brightness to our Sun, then whatever direction you look in you should always see a star. This would make the sky as bright as the Sun during the day and at night! This is obviously not the case but can be explained by the theory of the Big Bang and an expanding universe.

The universe is expanding at an accelerating rate; the light from the most distant galaxies will never reach us, as the space between us and those distant galaxies is always increasing faster than the time available for light to cross it. As the universe expands it also causes the light from distant galaxies to be redshifted towards wavelengths that we cannot see, e.g., infrared, microwaves. This is why the night sky appears dark.

TOP TIP

Measurements of the redshift of galaxies provides further evidence for the Big Bang.

Quick Test 21

1. Describe what is meant by the 'cosmic microwave background'.
2. What are the two most abundant elements in the universe?
3. Give two reasons why the night sky appears dark.

Orders of magnitude

The range of orders of magnitude

We express the order of magnitude of objects in powers of 10, i.e., an object that is 1000 times larger is three orders of magnitude bigger. Physics deals with quantities across a vast range of orders of magnitude, from the most minute of scales, such as those studied in particle physics, to the largest of scales necessary when studying vast distances in space.

The everyday world has an order of magnitude near the middle of this range. Human beings between 1 and 2 metres in height correspond to an order of magnitude of 10^0 m.

The table below gives examples of distances ranging from the human scale to the scale of the Earth.

1 m	Human scale – the average British person is 1·69 m
10 m	The height of a house
100 m	The width of a city square
10^3 m	The length of an average street
10^4 m	The diameter of a small city, such as Perth
10^5 m	The approximate distance between Aberdeen and Dundee
10^6 m	The length of Great Britain
10^7 m	The diameter of Earth

The diameter of the Earth is therefore 7 orders of magnitude bigger than a human being. To put things in perspective, we would need to repeat this increase four times over to reach the edge of the observable universe, at around 10^{28} m. In the other direction, we would need to repeat this change in size by more than twice the amount to reach the scale of the smallest particles that have been discovered.

Powers of 10

Size	Powers of 10	Examples
	10^{-18} m	Size of an electron/quark
1 fm (femto)	**10^{-15} m**	Size of a proton
	10^{-14} m	Atomic nucleus
1 pm (pico)	**10^{-12} m**	

Size	Powers of 10	Examples
	10^{-10} m	Atom
1 nm (nano)	10^{-9} m	Glucose molecule
	10^{-8} m	Size of DNA
	10^{-7} m	Wavelength of visible light
1 μm (micro)	10^{-6} m	Diameter of cell mitochondria
	10^{-5} m	Red blood cell
	10^{-4} m	Width of a human hair
1 mm (milli)	10^{-3} m	Width of a credit card
1 cm (centi)	10^{-2} m	Diameter of a pencil
	10^{-1} m	Diameter of a DVD
1 m	10^{0} m	Height of door handle
	10^{1} m	Width of a classroom
	10^{2} m	Length of a football pitch
1 km (kilo)	10^{3} m	Tallest building in the world – the BurjKalifa, Dubai
	10^{4} m	Cruising altitude of an aeroplane
	10^{5} m	Height of the atmosphere
1 Mm (mega)	10^{6} m	Length of Great Britain
	10^{7} m	Diameter of Earth
1 Gm (giga)	10^{9} m	Diameter of the Sun
	10^{11} m	Orbit of Venus around the Sun
1 Tm (tera)	10^{12} m	Orbit of Jupiter around the Sun
	10^{13} m	The edge of our solar system
	10^{16} m	Distance to nearest star Proxima Centauri
	10^{21} m	Diameter of our galaxy
	10^{23} m	Distance to the Andromeda galaxy
	10^{28} m	Distance to the edge of the observable universe

TOP TIP

You do not have to remember specific numbers to do with the size of particles or distances in space. You will, however, be required to make correct selections from data of relative sizes.

Quick Test 22

1. An object that is 1 million times larger is bigger by how many orders of magnitude?

2. In what area of physics do we find the smallest orders of magnitude?

3. Where do we find the largest scales in terms of magnitude?

The standard model

Matter and antimatter

The majority of ordinary matter consists of protons, neutrons and electrons. However, when particles are collided at high speed, many other particles can be produced. By passing these particles through a magnetic field, we can determine their mass and charge.

From the 1930s onwards, the advancement of particle accelerators has allowed physicists to develop an understanding of the fundamental nature of matter. The standard model of fundamental particles and interactions proposes 12 fundamental particles, organised into three generations.

High-energy collisions have also revealed the existence of antimatter. Every matter particle has an antiparticle. Antiparticles have the same mass as their matter counterparts, but have an opposite charge, e.g., an anti-electron or positron has the same mass as an electron, but has a positive charge.

Fermions

Fermions include protons, neutrons, electrons and their antimatter equivalents. All matter is made from fermions.

Electrons are fundamental particles, meaning they cannot be split up. Protons and neutrons are made up of other particles and so are not fundamental. These particles are called quarks. There are six types of quark, two of which make up protons and neutrons. A proton contains two up quarks and one down quark. A neutron contains one up quark and two down quarks.

The remaining quarks are: charm, strange, top and bottom. The electron belongs to a family of fundamental particles called leptons. There are three negatively charged leptons: the electron, the muon and the tau. Each of these charged leptons has an associated neutrino. Neutrinos have no charge.

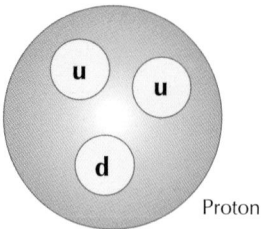 Proton

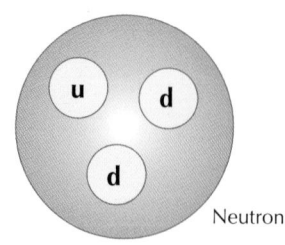 Neutron

The fundamental fermions can be summarised in the following table. There are three 'families' of fundamental particles:

		First generation	Second generation	Third generation
Quarks		up	charmed	top
		down	strange	bottom
Leptons		electron	muon	tau
		electron neutrino	muon neutrino	tau neutrino

Neutrinos are probably the most numerous particles in the universe but are very difficult to detect. The first evidence for the neutrino came from the emission of beta particles from the nuclei of radioactive atoms. It was found in the process of beta decay that energy was not conserved, so it was suggested that another particle, the neutrino, is also emitted. The emission of the neutrino means that energy is conserved during beta decay.

Hadrons, baryons and mesons

Hadrons are matter particles that consist of combinations of quarks. There are two different types of hadron. These are called baryons and mesons.

Baryons are made up of **three quarks**. Protons and neutrons and their antiparticles are baryons. A proton consists of two up quarks and one down quark. A neutron consists of one up quark and two down quarks.

Mesons are made up of **two quarks**, a quark and an anti-quark. All mesons are very unstable, and so have very short lifetimes.

Bosons

All interactions between matter particles are governed by forces. There are four fundamental forces in the universe: gravity, electromagnetism, the strong nuclear force, and the weak nuclear force.

Each force has a particle that is used to exchange force between particles when they interact. These particles are called bosons.

The photon is the exchange particle for the electromagnetic force. The gluon is the exchange particle for the strong nuclear force which acts between quarks to hold neutrons in the nucleus. W and Z bosons are the exchange particles for the weak nuclear force, the force that is responsible for radioactive decay.

TOP TIP

The gravitational force is thought to be carried by bosons called gravitons but these are yet to be detected experimentally.

Quick Test 23

1. What is an antiparticle?
2. How many fundamental particles are there in the standard model?
3. Describe the differences between baryons and mesons.
4. Name three bosons and their corresponding forces.

Force fields

Fields

A field can be thought of as a region where objects will experience a force. We can represent the strength of a field and the direction of force by using field lines.

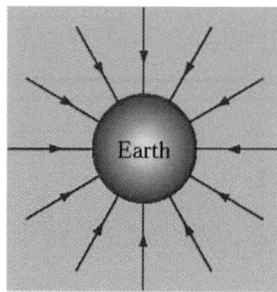

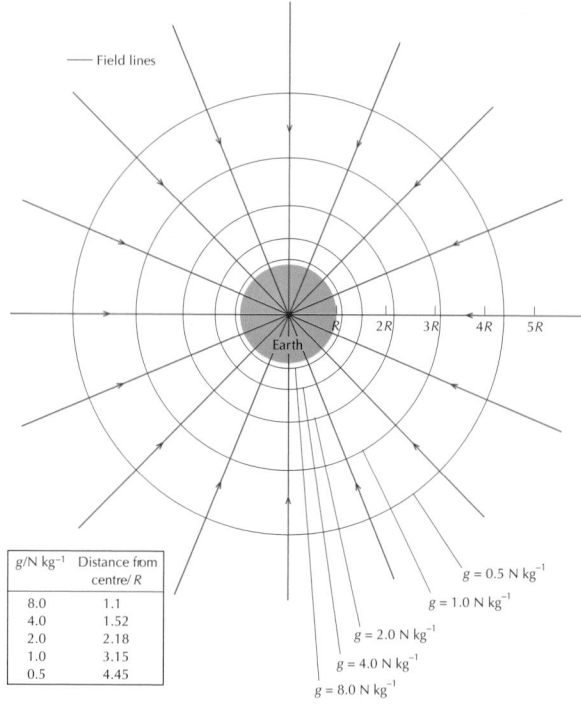

Field lines

g/N kg⁻¹	Distance from centre/ R
8.0	1.1
4.0	1.52
2.0	2.18
1.0	3.15
0.5	4.45

$g = 0.5$ N kg^{-1}
$g = 1.0$ N kg^{-1}
$g = 2.0$ N kg^{-1}
$g = 4.0$ N kg^{-1}
$g = 8.0$ N kg^{-1}

For example, a gravitational field exists around the Earth and any object with mass that enters that field will experience a force acting towards the Earth. In general the strength of the field reduces with distance, meaning that an object will experience less of a gravitational force the further it is from the Earth.

Electric fields operate in a similar way to gravitational fields. Gravitational fields exist around objects that have mass, in the same way that electric fields exist around objects that have charge. The size of the mass determines the strength of the gravitational field, just as the quantity of charge determines the strength of the electric field. The key difference is that gravitational fields are always attractive, whereas electric fields can be attractive and repulsive.

Electric charge

Electric charge has the symbol Q and is measured in coulombs (C). The charge on a proton is $1.6 \times 10^{-19} C$. The charge on an electron is the same magnitude but has the opposite sign, i.e., $-1.6 \times 10^{-19} C$. If there is an equal amount of positive and negative charge, then an object will not be surrounded by an electric field. There has to be an excess of one particular type of charge (positive or negative) in order for an electric field to exist around an object.

When a charged particle enters an electric field it will experience a force. The force that it experiences depends on the charges of the objects involved. Opposite charges attract one another, and like charges repel each other. The direction of an electric field is the direction of force that a positive charge experiences in the field. Therefore, electric field lines are always directed from positive to negative charges.

The separation of the field lines indicates the strength of the field. Much like gravitational fields, the strength of the field decreases as you move further away from the charge.

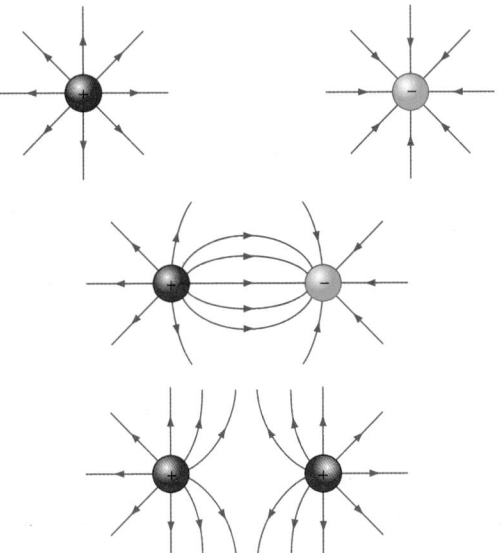

Uniform electric field

The field between two charged parallel plates is uniform. The field is therefore a constant strength at all points between the plates.

The uniform electric field is indicated by the equally spaced field lines.

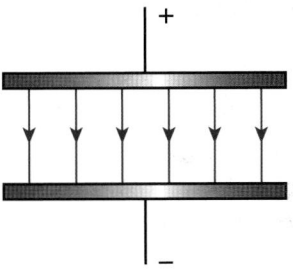

Quick Test 24

1. In physics, what is meant by a field?

2. In which direction do electric field lines 'flow'?

3. How do we use field lines to indicate a uniform electric field?

Electric fields

Movement of charge in an electric field

Consider a positive charge placed in a uniform electric field as shown.

A force is exerted on the charge. Because it is a positive charge, this force will be in the direction of the field lines, causing the positive charge to accelerate towards the negative plate. The positive charge gains kinetic energy as it travels from A to B.

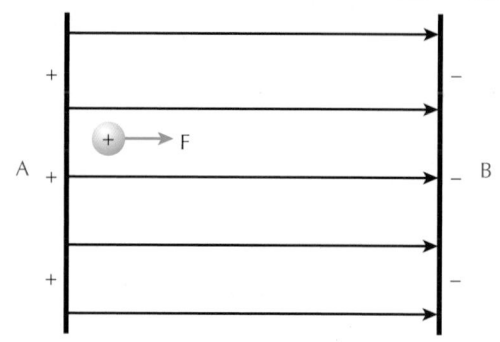

Electrical potential energy

If the charge goes from B to A, then energy will have to be supplied to overcome the forces of repulsion.

In moving from B to A, the charge gains electrical potential energy. The work done in moving the charge from B to A is equal to the gain in electrical potential energy.

This is a similar situation to objects in gravity fields. Work must be done against the force of gravity to raise an object from the ground. The work done is equal to the gain in gravitational potential energy. When the object is released, the potential energy is transferred to kinetic energy.

Definition of the volt

If it requires 1 joule of work to move a charge of 1 coulomb between two points, we say that the potential difference is 1 volt.

1 volt = 1 joule per coulomb

$1\,V = 1\,J\,C^{-1}$

This leads to the formula:

$$V = \frac{W}{Q}$$

where V is the potential difference in volts

Q is the charge in coulombs

W is the electrical potential energy or work done in joules

The equation given in the relationships sheet is:

$$W = QV$$

By conservation of energy, the electrical potential energy gained, W, is equal to the kinetic energy, E_K, of the charge when it is released.

$$W = E_k$$

$$QV = \frac{1}{2}mv^2$$

TOP TIP

This relationship is **not given** in the Higher relationships sheet.

Calculations involving electric fields

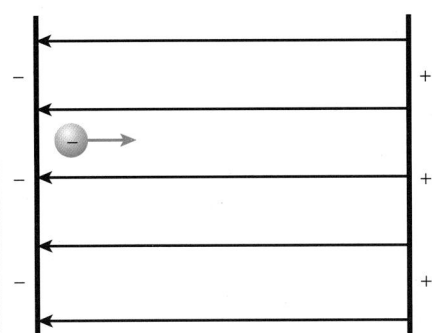

A typical electric flow problem is as follows:

An electron is moved between two parallel plates. The potential difference between the plates is 5kV.

Calculate the speed of the electron on reaching the positive plate.

First calculate the electrical potential energy lost in moving the electron between the two plates.

$$W = QV$$
$$W = 1 \cdot 6 \times 10^{-19} \times 5000$$
$$W = 8 \times 10^{-16}\,J$$

By conservation of energy, the electrical potential energy lost is equal to the kinetic energy gained as the electron moves towards the positive plate.

$$QV = \frac{1}{2}mv^2$$

$$8 \times 10^{-16} = \frac{1}{2} \times 9 \cdot 11 \times 10^{-31} \times v^2$$

$$v = \sqrt{\frac{8 \times 10^{-16}}{4 \cdot 055 \times 10^{-31}}}$$

$$v = 4 \cdot 4 \times 10^7\,m\,s^{-1}$$

Quick Test 25

1. What type(s) of energy is/are gained or lost by a charge as it moves through an electric field?

2. Define what is meant by a potential difference of 1 volt.

3. A proton is accelerated from rest by a potential difference of 2kV.
   Calculate the speed of the proton (mass of proton = $1 \cdot 673 \times 10^{-27}$ kg).

Moving charges and magnetic fields

Magnetic fields

Magnets are surrounded by magnetic fields. This creates a force on other magnets and certain other materials that enter the field.

The Earth has a magnetic field that can be represented with field lines in a similar way to gravitational and electric fields. Field lines for magnetic fields always run from north to south.

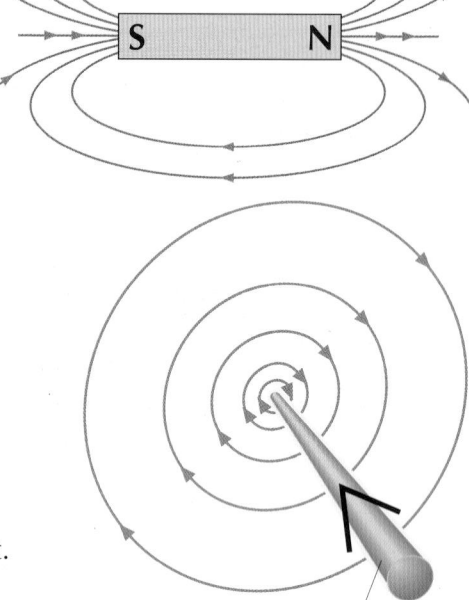

Moving charges create magnetic fields. For example, a current carrying wire creates a magnetic field around it. The magnetic field around the wire is circular.

Wire carrying current

Moving charges in a magnetic field

When two magnets interact, they either attract or repel each other because of the magnetic fields surrounding each magnet. A moving charge creates its own magnetic field so when it passes through an external magnetic field it experiences a force. The direction of this force depends on the type of charge (positive or negative), and the direction of the magnetic field through which it passes.

Forces acting on moving charges in a magnetic field

The direction of force acting on a charge moving through a magnetic field can be predicted by two simple rules. There is one rule for positive charges moving in a magnetic field and another rule for negative charges moving in a magnetic field.

Movement of positive charge in a magnetic field

The left-hand rule gives the direction of force for a positive charge moving in a magnetic field. The first finger must point in the direction of the magnetic field. The second finger must point in the direction of conventional current flow (+ to –). The thumb indicates the direction of force. Note that the force is at right angles to the magnetic field and the current is at right angles to both the magnetic field and the force.

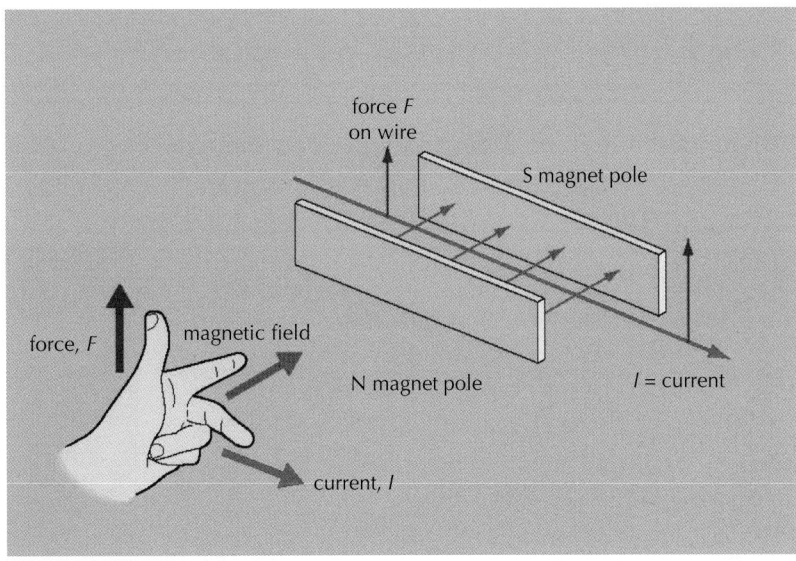

Movement of negative charge in a magnetic field

The right-hand rule gives the direction of force for a negative charge moving in a magnetic field. The positioning of the fingers is identical to that of the left-hand rule. This time though, the second finger indicates direction of electron flow current (– to +).

> **TOP TIP**
>
> The direction of the current flow can be taken as the direction of the moving charge.

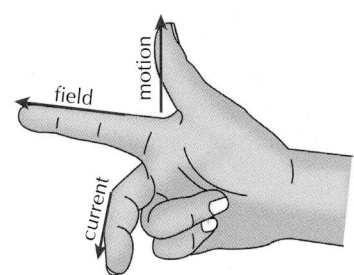

Quick Test 26

1.   What creates magnetic fields?
2.   Which rule should be used for positive charges moving in a magnetic field?
3.   Which rule should be used for negative charges moving in a magnetic field?

Particle accelerators

Types of particle accelerator

There are three main types of particle accelerator:

- linear accelerators
- cyclotrons
- synchrotrons

In a linear accelerator, particles are accelerated in a straight line towards a target using only electric fields. A combination of electric and magnetic fields are used to accelerate particles in a circular path in cyclotrons and synchrotrons – like the Large Hadron Collider (LHC).

The basic operation of all particle accelerators is the same.

Acceleration

The purpose of particle accelerators is to produce high-energy collisions at high speeds. To achieve this, particles are accelerated by rapidly changing electric fields. Through very rapid voltage switching, the particles can be attracted or repelled by the electric field. If this switching is timed correctly then the particles can reach speeds close to the speed of light.

Deflection

In cyclotrons and synchrotrons, the accelerated particles move in a circular path. The particles would move in a straight line, unless acted upon by an external force. This external force comes from powerful magnets positioned around the accelerator. There are over 9000 superconducting magnets in the LHC, which operate at temperatures close to absolute zero.

The magnetic field is used to deflect the particles into a path of constant radius. As the particles travel faster, the strength of the magnetic field is increased. When the particles have the desired kinetic energy, the magnets are used to bring about particle collisions in a desired location.

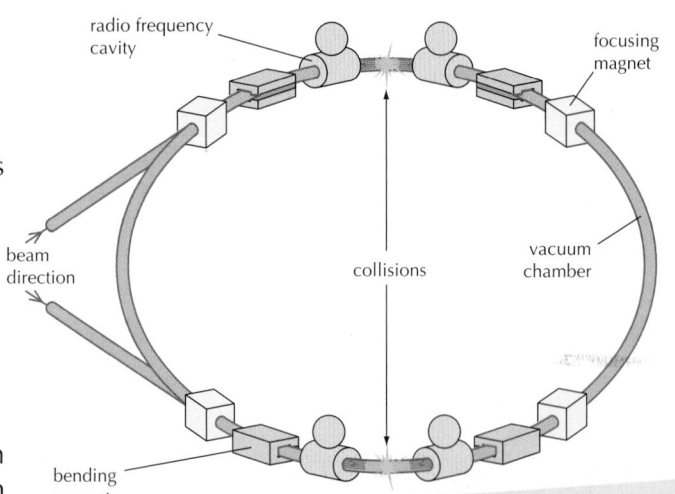

TOP TIP
Particle accelerators use a combination of electric and magnetic fields to create high-energy collisions.

Collision

In a linear accelerator the particles are collided with a stationary target. In the case of the LHC, two proton beams travelling in opposite directions are collided together at four possible detector locations. In both cases, a very high number of collision events are produced.

The large amounts of data are then analysed using highly sophisticated computer programs to identify the properties of the particles produced in the collisions. Often the particles themselves are not directly observed but the effects of them can be detected, as was the case with the discovery of the Higgs Boson in 2012. One of the detectors, ATLAS, used in the discovery of the Higgs Boson is shown below:

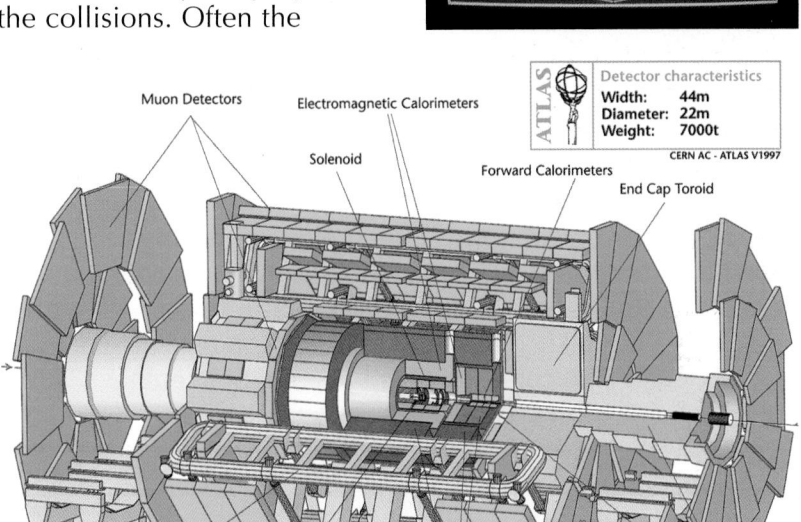

	Detector characteristics
ATLAS	Width: 44m Diameter: 22m Weight: 7000t

CERN AC - ATLAS V1997

The particles are detected at different layers in the system:

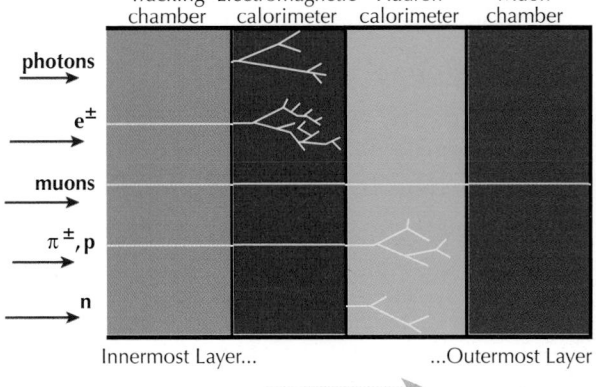

Quick Test 27

1. How do particles reach high speeds in particle accelerators?
2. What keeps particles moving in circular paths in cyclotrons and synchrotrons?

Model of the atom

A basic model of the atom

When studying nuclear reactions, we are dealing with energy released from the nucleus of an atom. A basic model of the atom consists of a nucleus of protons with mass number 1 and charge +1, and neutrons with mass number 1 and charge 0.

The nucleus of an atom is represented as $^A_Z X$

Z is the number of protons in the nucleus and is called the **atomic number**.

A is the total number of protons and neutrons in the nucleus and is called the **mass number**.

The element is determined by Z, the number of protons. In a neutral atom, the number of protons in the nucleus is equal to the number of electrons in orbit. Different atoms of the same element can have different values of A, due to having more or fewer neutrons. These different versions of the same element are called **isotopes**. An isotope has the same number of protons but a different number of neutrons, i.e., it has the same atomic number but a different mass number.

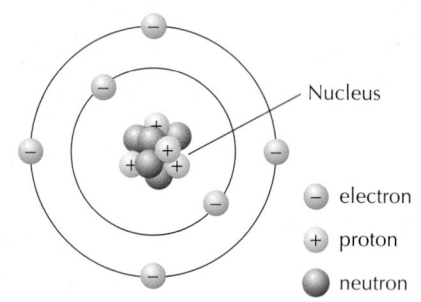

Nucleus

− electron
+ proton
● neutron

Radioactive decay

Radioactive decay is the breakdown of a nucleus to release energy and matter from that nucleus. Nuclei will only release energy or matter if they are unstable. Unstable nuclei are called radioisotopes. The original nucleus is called the parent nucleus. The product of the radioactive decay is called the daughter nuclei.

Alpha decay

In alpha decay, an alpha particle which has two protons and two neutrons is spontaneously emitted from a nucleus. Alpha decay reduces the atomic number of the parent nucleus by 2 and reduces the mass number by 4.

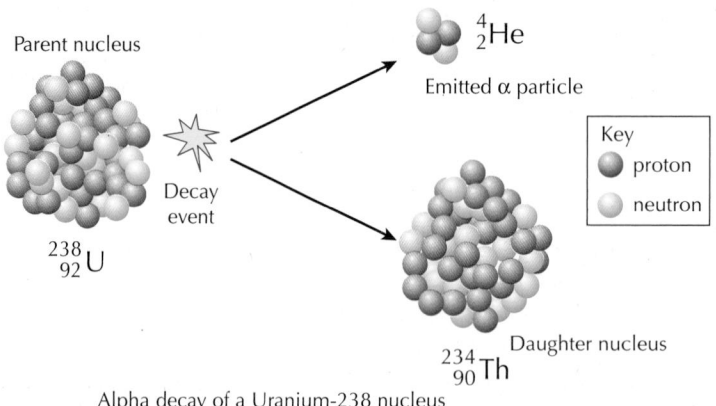

Parent nucleus

$^{238}_{92}U$

Decay event

^4_2He

Emitted α particle

Key
● proton
○ neutron

$^{234}_{90}Th$

Daughter nucleus

Alpha decay of a Uranium-238 nucleus

Beta decay

In beta decay, a fast moving electron is emitted from the nucleus. It has the symbol $_{-1}^{0}e$ or $_{-1}^{0}\beta$

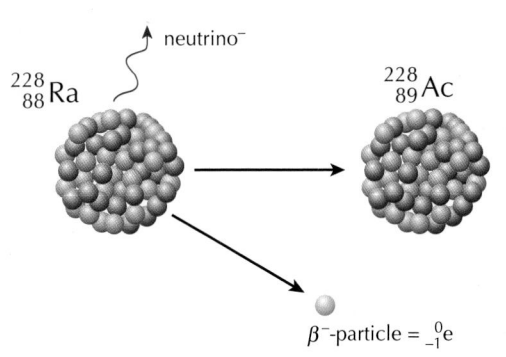

A neutron in the nucleus changes into a proton (due to an up quark changing into a down quark) with the emission of an electron and an antineutrino, $\bar{\nu}$. This causes the atomic number to increase by 1 but the mass number remains unchanged.

For example, the beta decay of Carbon-14:

$$_{6}^{14}C \rightarrow _{7}^{14}N + _{-1}^{0}e + _{0}^{0}\bar{\nu}$$

Gamma decay

Alpha and beta radiation are often accompanied with the emission of gamma radiation. Gamma rays are a type of electromagnetic radiation that can be produced due to a redistribution of charge in the nucleus. Gamma radiation is often emitted following alpha and beta emission, as the nucleus is left in a higher-energy, excited state. Emission does not affect the mass and atomic number, as gamma is an electromagnetic wave, and not a particle.

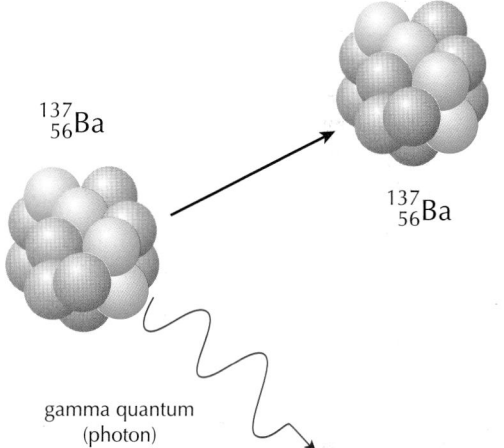

gamma quantum (photon)

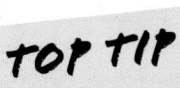

TOP TIP

Alpha, beta and gamma radiation are often identified by the corresponding Greek letter. Alpha (α); beta (β); gamma (γ).

Quick Test 28

1. An element can be represented $_{Z}^{A}X$. Identify the meaning of the letters A and Z.

2. What is meant by an isotope?

3. What effect does alpha decay have on the parent nucleus?

4. What effect does beta decay have on the parent nucleus?

Nuclear fission

Spontaneous decay

Nuclear fission is the process of a heavy nucleus splitting to form two lighter nuclei. This radioactive decay is called **spontaneous fission**. During the decay process, free neutrons are produced and energy is released. The neutrons released by spontaneous fission can cause a chain reaction, in which more nuclei split as they absorb the released neutrons.

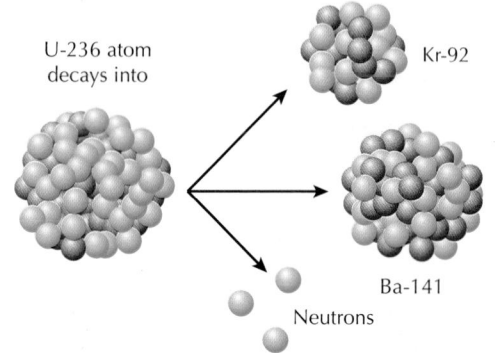

U-236 atom decays into

Kr-92

Ba-141

Neutrons

Neutron bombardment

Fission can occur as the result of neutron bombardment. This is known as **induced fission**. Induced fission takes place when a large nucleus absorbs a neutron.

When Uranium-235 absorbs a neutron as shown above, it undergoes fission. This can be represented in the equation:

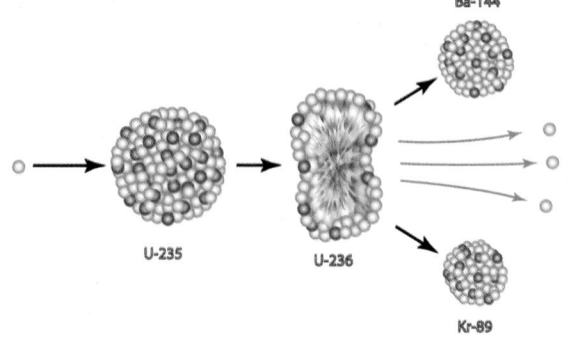

Ba-144

U-235

U-236

Kr-89

$$^{235}_{92}U + ^{1}_{0}n \rightarrow ^{144}_{56}Ba + ^{89}_{36}Kr + 3\,^{1}_{0}n + energy$$

Notice that mass number and atomic number are both conserved during the fission reaction.

Nuclear fission and E = mc²

Although mass number and atomic number are both conserved in fission reactions, if we accurately measure the mass before and after fission, we find that there is a mass difference. The total mass before fission is greater than the total mass after the fission process. This mass difference is called a **mass defect**.

The loss of mass is accompanied by an increase in energy. Einstein linked the loss of mass with the increase in energy using the equation:

> **$E = mc^2$**
>
> E is the energy released in joules (J)
> m is the mass defect measured in kilograms (kg)
> c is the speed of light in a vacuum (m s^{-1})

Example

A fission reaction is represented by the following equation:

$$^{235}_{92}U + ^{1}_{0}n \rightarrow ^{134}_{52}Te + ^{98}_{40}Zr + 4\,^{1}_{0}n$$

The energy produced in the reaction can be calculated by first comparing the total mass before and after fission.

Mass before:

Uranium	$3{\cdot}901 \times 10^{-25}$ kg
Neutron	$0{\cdot}017 \times 10^{-25}$ kg
Total	$3{\cdot}918 \times 10^{-25}$ kg

Mass after:

Tellurium	$2{\cdot}221 \times 10^{-25}$ kg
Zirconium	$1{\cdot}626 \times 10^{-25}$ kg
4 × neutrons	$0{\cdot}068 \times 10^{-25}$ kg
Total	$3{\cdot}915 \times 10^{-25}$ kg

Next calculate the mass defect: $3{\cdot}918 \times 10^{-25} - 3{\cdot}915 \times 10^{-25} = 0{\cdot}003 \times 10^{-25}$ kg

Energy released:

$E = mc^2$

$E = 0{\cdot}003 \times 10^{-25}(3 \times 10^{8})^2$

$E = 2{\cdot}7 \times 10^{-11} J$

> ### TOP TIP
> Be careful when rounding numbers during mass defect calculations. The mass defect will always be very small, so any intermediate rounding can vastly affect your answer.

Nuclear fission in nuclear reactors

Nuclear fission reactions release large amounts of heat energy. A nuclear reactor provides a means of harnessing the fission reactions in order to produce new forms of more useful energy.

The moderator is used to slow down the neutrons so they are more likely to cause further fission. The control rods absorb neutrons to control the chain reaction. The coolant is used to transfer the heat energy from the reactor to turn water into steam. This steam is then used to drive electrical generators to produce electricity.

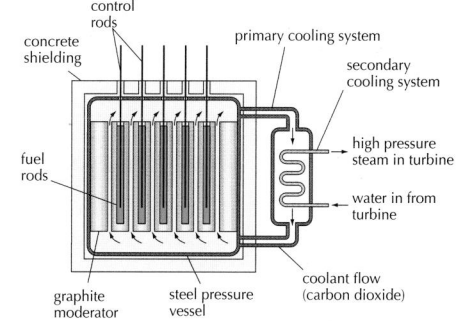

Fission reactors are contained within reinforced concrete, and lead to reduced contamination of the surroundings.

Quick Test 29

1. Describe the process of nuclear fission.

2. What is meant by the mass defect in a fission reaction?

3. Give one safety aspect of nuclear fission reactors.

Nuclear fusion

Fusion reactions

Nuclear fusion is the process of combining two lighter nuclei to form a nucleus with a larger mass number.

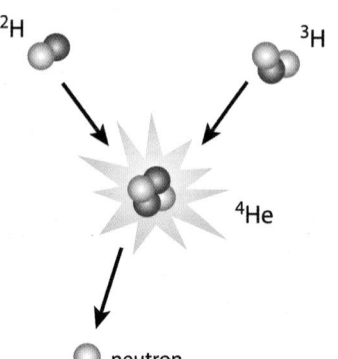

The reaction shown above can be written in the following equation:

$$^{2}_{1}H + ^{3}_{1}H \rightarrow ^{4}_{2}He + ^{1}_{0}n + energy$$

Notice that the mass number and atomic number are both conserved during the fusion reaction. When the nuclei combine they release energy. This type of reaction takes place in the sun and also in a hydrogen bomb.

TOP TIP

Hydrogen-2 is often called 'deuterium', with hydrogen-3 being 'tritium'.

Nuclear fusion and E = mc²

When nuclei fuse together, the total mass after the fusion reaction is less than the total mass before the fusion reaction. This mass difference is called the **mass defect**.

The loss of mass is accompanied by an increase in energy. Einstein linked the loss of mass with the increase in energy using the equation:

$E = mc^2$

E is the energy released in joules (J)

m is the mass defect measured in kilograms (kg)

c is the speed of light in a vacuum (m s⁻¹)

Example

A fusion reaction is represented by the following equation:

$$^{2}_{1}H + ^{2}_{1}H \rightarrow ^{3}_{1}H + ^{1}_{1}H$$

The energy produced in the reaction can be calculated by first comparing the total mass before and after fusion.

Mass before:

Deuterium	$3·342 \times 10^{-27}$ kg
Deuterium	$3·342 \times 10^{-27}$ kg
Total	$6·684 \times 10^{-27}$ kg

Mass after:

Tritium	$5·005 \times 10^{-27}$ kg
Hydrogen	$1·672 \times 10^{-27}$ kg
Total	$6·677 \times 10^{-27}$ kg

Next calculate the mass defect: $6·684 \times 10^{-27}$ kg $- 6·677 \times 10^{-27}$ kg $= 7 \times 10^{-30}$ kg

Energy released:

$E = mc^2$

$E = 7 \times 10^{-30}(3 \times 10^8)^2$

$E = 6·3 \times 10^{-13}\,J$

Nuclear fusion reactors

A controlled fusion reaction can be achieved in a device known as a Tokamak. Hydrogen molecules are brought to extremely high temperatures of over 100 million kelvin. At these temperatures, the hydrogen gas molecules separate into atoms, then into electrons and nuclei, forming a plasma.

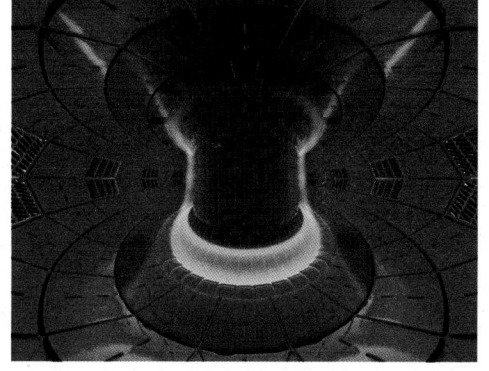

At these high temperatures, the ordinary solid matter of a container would melt or evaporate. Strong magnetic fields are used to confine the plasma and keep it moving in circles in the shape of a hollow doughnut or **torus**. The magnetic field also keeps the plasma away the edges of the container.

Quick Test 30

1. Describe the process of nuclear fusion.

2. How does the mass after a fusion reaction compare with the mass before the fusion reaction?

3. Describe one safety issue with nuclear fusion reactors.

The photoelectric effect

Photoemission

Sometimes, when electromagnetic radiation strikes a metal surface electrons are emitted. This is the basis of solar cells and LDRs.

When this phenomenon was first discovered at the end of the 19th century, the experimental results could not be explained by the wave theory of light.

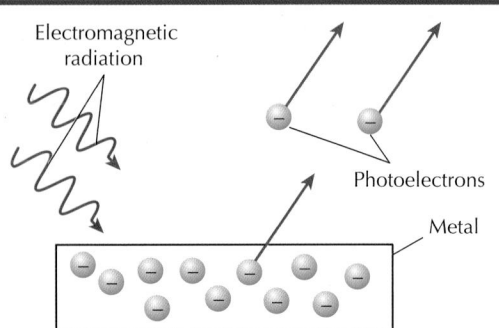

The gold-leaf electroscope

Static electricity can be used to give a gold-leaf electroscope a negative charge. The leaf will only rise if the electroscope is charged.

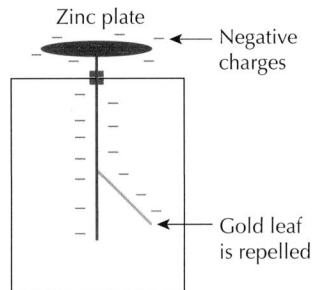

> **TOP TIP**
>
> The leaf rises because of mutual repulsion between the negative charges.

Electromagnetic radiation can be directed at the metal plate. In most circumstances, there is no effect when electromagnetic radiation strikes the metal plate. For example, white light of varying irradiance is shone on the zinc metal plate, but no effect is observed. The emission of electrons will only occur for certain high frequencies of electromagnetic radiation. Photoemission also depends on the type of metal, e.g., ultraviolet radiation would cause zinc to discharge, but ultraviolet radiation would have no effect on a chromium plate.

We can tell that photoelectrons are being emitted from the negatively charged electroscope as the leaf begins to fall. The more intense the ultraviolet radiation, the more electrons are ejected. This photoelectric effect cannot be explained by thinking of light as a continuous wave. To eject an electron requires a precise amount of energy, the value of which depends on the frequency of the radiation. These discrete packets of energy are called photons.

> **TOP TIP**
>
> If photoemission does not occur for a particular frequency of radiation, increasing the irradiance will have no effect.

Photons

In order to explain the photoelectric effect, Albert Einstein and Max Planck proposed the idea that light was not a continuous wave, but existed as a stream of 'packets' or 'quanta'. These quanta are called photons, and are particles of electromagnetic radiation that have zero mass. Each photon has a wavelength and frequency associated with it.

A lamp emits random bursts of energy. Each burst is a photon, a quantum of radiation.

Thinking of light as a particle rather than a wave helps to explain the photoelectric effect. White light does not eject electrons from zinc or chromium, regardless of the irradiance (see page 74). If light was acting as a continuous wave then the energy delivered to the metal plate should eventually be enough to cause photoemission. In practice, this never occurs. When ultraviolet radiation strikes the zinc metal plate, the electrons are ejected because each photon has enough energy to eject to do so.

Wave-particle duality

The wave particle duality principle states that matter and light exhibit the behaviours of both waves and particles. For example light will, in some situations, behave as a particle and in others, as a wave. The phenomenon of interference (see page 68) can only be explained by the wave theory of light. The photoelectric effect can only be explained by the particle theory of light.

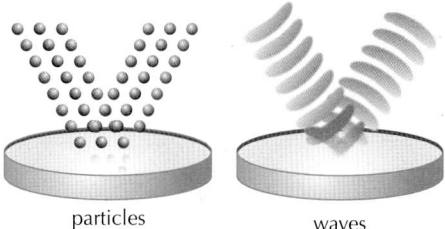

particles   waves

Although photons are treated as particles, they still exhibit wave-like properties. For example, photons have frequency, wavelength and amplitude.

Quick Test 31

1. What is photoemission?
2. What factors determine whether or not photoemission takes place?
3. What causes the leaf on the electroscope to fall?
4. Research situations where light behaves as a wave and a particle. Find at least two examples of each.

Quantum theory of light

Threshold frequency and work function

Every metal has its own particular frequency required to eject an electron from it. The minimum frequency required for photoemission to take place is called the threshold frequency.

Photons at or above this threshold frequency will eject photoelectrons; photons below this frequency will not eject photoelectrons.

The energy contained in each photon is given by:

$$E = hf$$

where E is the energy in joules (J)

f is the frequency in hertz (Hz)

h is Planck's constant

The energy of the photon is directly proportional to its frequency. Some metals have electrons that are more tightly bound to their atoms. These metals therefore require more energy to eject an electron.

The minimum energy required for photoemission is called the work function.

$$work\ function = hf_o$$

where f_o is the threshold frequency

If a photon carries more energy than the work function, the electron is not only ejected, but the extra energy left over becomes kinetic energy for the escaping photoelectron.

$$Kinetic\ energy\ of\ electron = E_{photon} - work\ function$$

$$E_k = hf - hf_o$$

The kinetic energy depends on the frequency of the radiation, not the irradiance (see page 74.) Higher irradiance produces more electrons, but does not increase the kinetic energy or speed of the electrons. Irradiance increase simply means more photons arriving, *not* an increase on the energy of each photon. Energy increase is only achieved through an increase in frequency.

Worked example

A metal surface has a work function of $2 \cdot 87 \times 10^{-19}$ J. Light of wavelength 588 nm is incident on the metal surface.

What is the maximum kinetic energy with which electrons will be emitted from the surface?

Photon energy is directly proportional to frequency, so calculate the frequency of the radiation first:

$$v = f\lambda$$

$$3 \times 10^8 = 588 \times 10^{-9} f$$

$$f = 5 \cdot 1 \times 10^{14} Hz$$

Next calculate the energy of the photon:

$$E = hf$$

$$E = 6 \cdot 63 \times 10^{-34} \times 5 \cdot 1 \times 10^{14}$$

$$E = 3 \cdot 38 \times 10^{-19} J$$

The maximum kinetic energy is equal to the difference between the energy of the photon and the work function:

$$E_k = hf - hf_0$$

$$E_k = 3 \cdot 38 \times 10^{-19} - 2 \cdot 87 \times 10^{-19}$$

$$\underline{E_k = 5 \cdot 13 \times 10^{-20} J}$$

Quick Test 32

1. What is meant by the threshold frequency?
2. What is the work function of a metal?
3. What determines the maximum kinetic energy of a photoelectron?

Interference

Constructive and destructive interference

Interference occurs when waves from two or more sources meet. When waves meet, the resultant wave depends on the amplitude and the relative phase of the waves.

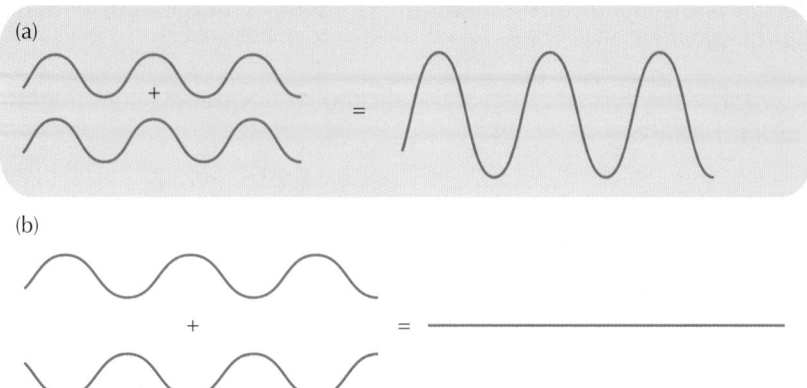

If two waves meet exactly in phase (crest meets crest, or trough meets trough), they will interfere constructively to form a wave of greater amplitude.

If two waves meet exactly out of phase (crest meets trough), they will interfere destructively to form a wave of reduced amplitude. If both waves are of the same amplitude then they will cancel each other out completely.

TOP TIP

When asked to describe what is meant by constructive or destructive interference, make sure to use the word **'meet'**, e.g., the waves **meet** in phase and interfere constructively.

Coherent waves

In order to observe a steady interference pattern, the wave sources need to have a fixed phase difference. In order to achieve this, the waves must also have the same frequency, wavelength and velocity. Such wave sources are known as **coherent**.

Two identical loudspeakers emitting waves of the same wavelength amplitude and velocity, with a constant phase relationship, will produce coherent sound waves.

If an observer walks slowly from one speaker to the other, they will detect alternating loud and quiet regions.

The loud regions are where the two waves meet in phase and interfere constructively. This area of constructive interference is called a **maxima**.

The quiet regions are where the two waves meet out of phase and interfere destructively. This area of destructive interference is called a **minima**.

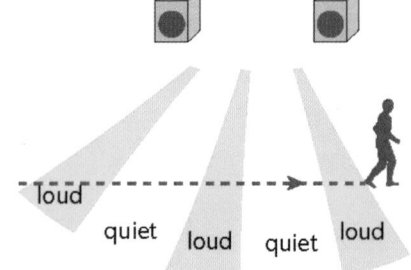

Path difference

If two wave sources are coherent, then at any point at the same distance from each source, the waves will arrive in phase, interfering constructively. This is shown at point O in the diagram where the distance AO is 3 wavelengths and the distance BO is also 3 wavelengths. If the wave from one source has travelled further than the other, there is a path difference between the waves. If this path difference is a whole number of wavelengths, a maxima will be produced. If the path difference includes one half of a wavelength, a minima will be produced. This is shown at point P in the diagram where the distance AP is 2.5 wavelengths and the distance BP is 3 wavelengths.

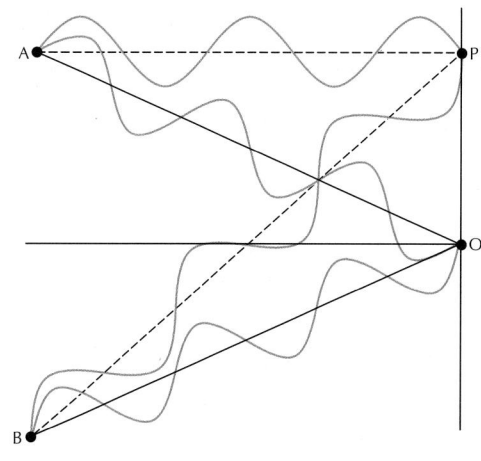

For constructive interference:

path difference $= m\lambda$ where m = 0, 1, 2 ...

For destructive interference:

path difference $= \left(m + \frac{1}{2}\lambda\right)$ where m = 0, 1, 2 ...

Example

A source of microwaves is placed in front of a metal sheet that has two slits S_1 and S_2.

A third order maximum is detected at point P.

Measurements of distances S_1P and S_2P are: $S_1P = 0\cdot421$ m and $S_2P = 0\cdot466$ m.

Calculate the wavelength of the microwaves.

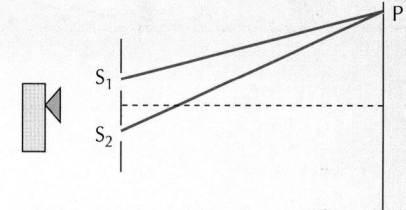

path difference $= S_2P - S_1P$

path difference $= 0\cdot466 - 0\cdot421$

path difference $= 0\cdot045\ m$

For a third order maximum:

path difference $= m\lambda$

$0\cdot045 = 3\lambda$

$\lambda = 0\cdot015\ m$

Quick Test 33

1. What is meant by constructive interference?
2. What is meant by destructive interference?
3. Describe the conditions required for two wave sources to be coherent.
4. What is meant by path difference?

Diffraction

Young's double slit experiment

It is difficult to produce coherent light sources because the emissions of light are random, and only last for a very short time. It is therefore difficult to produce a steady interference pattern.

Thomas Young was the first to devise an experiment to produce an interference pattern for light waves. He achieved this by passing light through a narrow slit, before using two slits to divide the wavefront. Passing the light through a single slit first ensures that the light waves are coherent. An interference pattern of bright and dark fringes is produced on a screen where the waves meet.

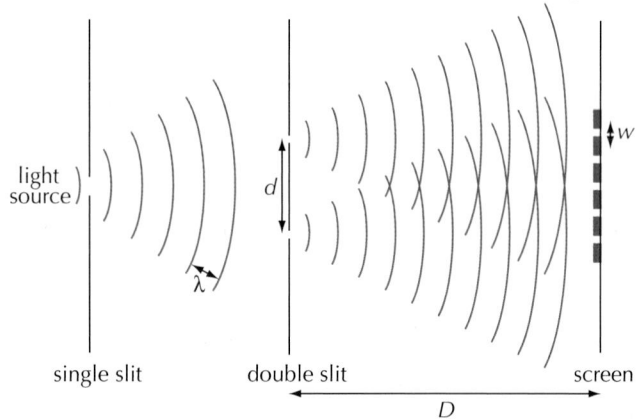

The bright fringes (maxima) occur when the path difference is a whole number of wavelengths, resulting in constructive interference. The dark fringes (minima) occur when the path difference is an odd number of half wavelengths resulting in destructive interference.

The distance between maxima or minima depends on:

- The distance between the two slits, d. Increasing d gives fringes that are closer together; decreasing d gives fringes that are further apart.

- The wavelength of the light, λ. The fringes are closer together for shorter wavelengths than for longer wavelengths.

- The distance, D, between the slits and the screen. The further away the screen is from the slits, the further apart the fringes will be.

Diffraction gratings

A diffraction grating is a series of narrow, parallel slits usually etched onto glass. Using more slits lets more light through and produces a series of bright sharp lines.

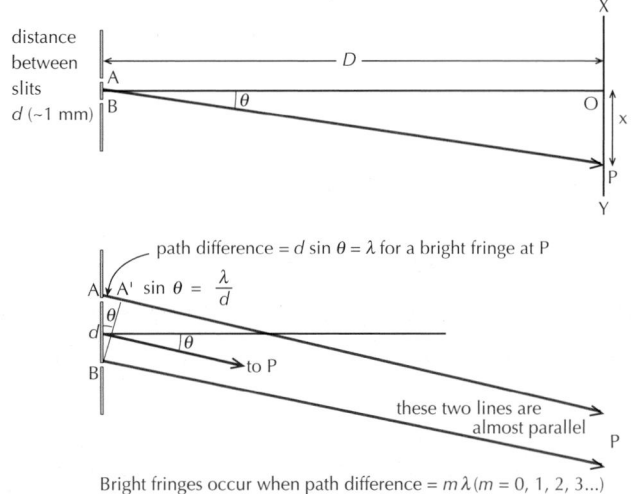

The path difference shown is AP − BP which is equivalent to $d\,sin\theta$. As with Young's double slit experiment, if the path difference is one wavelength, then a maxima will be produced.

Therefore:

$$d\,sin\theta = m\lambda$$

Bright fringes occur when path difference = $m\lambda$ (m = 0, 1, 2, 3...)

The first maximum will occur when m = 0, i.e., when there is no path difference. This is referred to as the zero order maximum. When m = 1 the path difference is exactly one wavelength. This is called the first order maximum.

Light of a single wavelength or frequency has a single colour and is called monochromatic light. When monochromatic light is passed through a diffraction grating, a series of bright fringes of one colour (e.g., red) will be produced.

Since the order of the maximum and, subsequently, the angle of deviation (θ), depend on the wavelength of the light, different colours of light will be produced at different angles. This means a diffraction grating can be used to disperse white light into the colours of the visible spectrum. At m = 0 where there is no path difference, a central white maximum will be produced. The first order maximum will show a spectrum of colour with violet light (the shortest wavelength) at the smallest angle, and red light (the longest wavelength) at the largest angle.

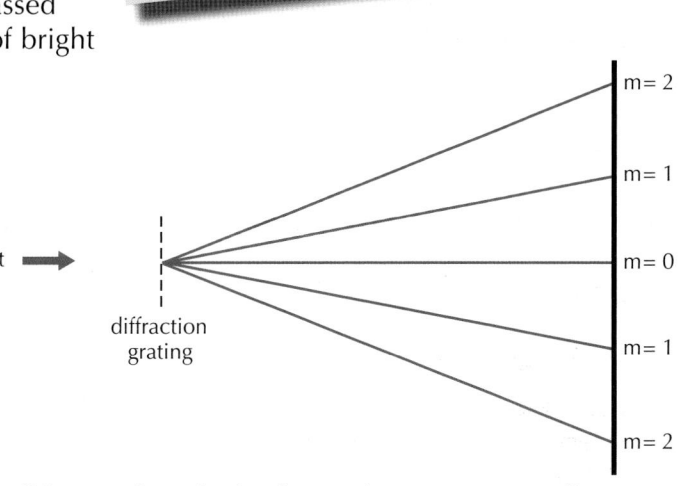

Quick Test 34

1. What should the path difference be between waves to produce maxima and minima?

2. Identify the variables that determine the spacing between the maxima and minima.

3. Calculate the spacing between the slits, *d*, for a diffraction grating of 100 lines per millimetre.

4. Light of wavelength 600 nm is shone onto a diffraction grating of 400 000 lines per metre. Calculate the angle between the zero order maximum and the first order maximum.

Snell's law

Refraction

When a wave moves from one medium to another, the wave changes speed. A good illustrative example of this is a light wave moving from air into glass. The change in speed can lead to a change in direction. The change in direction when a wave moves from one medium to another is called refraction.

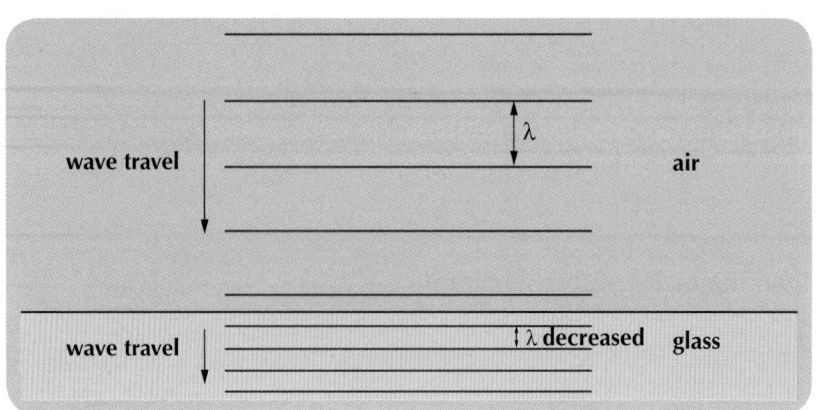

As the light wave enters the glass, its speed decreases. The wavefronts also become closer together, indicating that the wavelength has decreased. The frequency of the light remains the same throughout.

If the wavefront hits the boundary between the two media at an angle then the direction of the wave also changes. The greater the change in speed, the greater the change of direction. When moving from a more dense medium to a less dense medium (e.g., glass to air), the ray bends away from the normal line. When moving from a less dense medium to a more dense medium (e.g., air to glass), the ray bends towards the normal line.

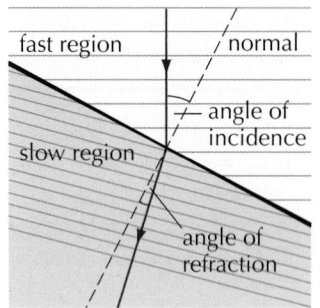

Refractive index

The refractive index of a medium is a measure of how much the material changes the speed of light as it passes through it. It therefore also gives a measure of how much the direction of the light changes as it passes from one medium to another.

The absolute refractive index of a material, n, is the refractive index of that material compared to the refractive index of a vacuum. The absolute refractive index of a vacuum (and also air) is $1 \cdot 00$.

Calculations with refraction

$n_1 \sin\theta_1 = n_2 \sin\theta_2$

because $n_1 = 1\cdot 00$, this simplifies to:

$\sin\theta_1 = n_2 \sin\theta_2$

$n_2 = \dfrac{\sin\theta_1}{\sin\theta_2}$

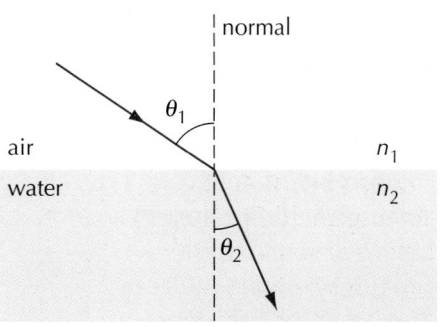

normal

air $\quad n_1$

water $\quad n_2$

θ_1

θ_2

The refractive index can also be calculated from the ratio of the speed of light in the two media, or by the ratio of the wavelengths in the two media:

$$n_2 = \frac{v_1}{v_2}$$

Given that $v = f\lambda$ and frequency is constant, the equation can be extended further to:

$$n_2 = \frac{\lambda_1}{\lambda_2}$$

TOP TIP

When calculating the ratio of $\dfrac{\sin\theta_1}{\sin\theta_2}$, $\dfrac{v_1}{v_2}$ or $\dfrac{\lambda_1}{\lambda_2}$ then the less dense material (usually air) is always on the top line of the equation.

As the refractive index is only a ratio, it does not have any units. The refractive index of all media is greater than $1\cdot 00$.

Different colours of light have different refractive indexes. Violet light has a larger refractive index than red light. Therefore, shorter wavelengths are refracted more than longer wavelengths. This is why a prism can be used to produce a visible light spectrum from white light.

Quick Test 35

1. State what is meant by the term refraction.

2. Define the term 'absolute refractive index'.

3. Give the three ways that the refractive index of a material can be determined.

4. A ray of blue light passes from air into glass. The refractive index of the glass for blue light is 1·65. Calculate the speed of light in the glass.

Total internal reflection

Critical angle

When a ray of light passes from a more dense medium to a less dense medium, there is a particular angle of incidence that results in an angle of refraction of 90°. This angle of incidence is called the critical angle, θ_c.

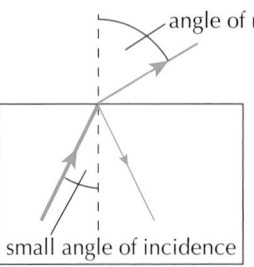

small angle of incidence

there is a weak reflection

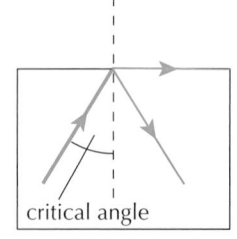

critical angle

there is a stronger reflection

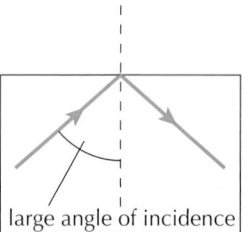

large angle of incidence

all light is reflected

For angles of incidence less than the critical angle, some light will be faintly reflected in addition to the refracted ray. At the critical angle there will be a much stronger reflection to accompany the refracted ray at 90°. For angles of incidence that are greater than the critical angle, all light will be reflected inside the material. This effect is called total internal reflection.

$$n = \frac{\sin\theta_a}{\sin\theta_c}$$

$$n = \frac{\sin 90}{\sin\theta_c}$$

$$n = \frac{1}{\sin\theta_c}$$

$$\sin\theta_c = \frac{1}{n}$$

TOP TIP

No refraction takes place at angles of incidence greater than the critical angle.

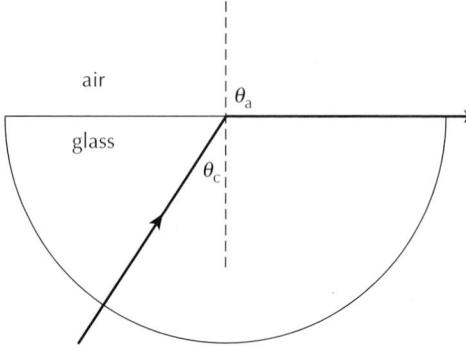

Uses of internal reflection

Total internal reflection allows signals to be transmitted across large distances, through optical fibres made from glass or plastic. If the light hits the boundary of the fibre at an angle greater than the critical angle, repeated internal reflection will occur and no light will pass out of the fibre. This method can be used to send television, radio and internet signals. Optical fibres are also used in endoscopy to look inside a patient's body. The flexibility of the optical fibres means that only a small incision is required. This type of 'keyhole surgery' has a much shorter recovery time than open surgery.

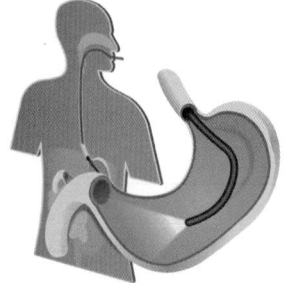

Worked example

A ray of blue light is directed at a glass prism as shown.

The refractive index of the glass for the blue light is 1·64.

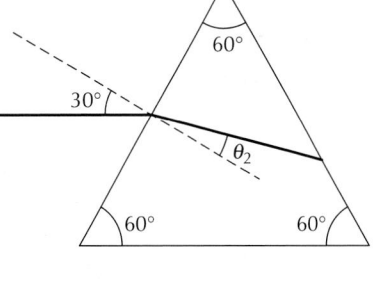

(a) Show that the angle θ_2 is 17·8°.

$$n = \frac{\sin\theta_1}{\sin\theta_2}$$

$$1\cdot 64 = \frac{\sin 30}{\sin\theta_2}$$

$$\theta_2 = 17\cdot 8°$$

(b) Calculate the critical angle for the blue light in the prism.

$$\sin\theta_c = \frac{1}{n}$$

$$\sin\theta_c = \frac{1}{1\cdot 64}$$

$$\sin\theta_c = 37\cdot 6°$$

(c) Sketch a diagram of the path of the blue light until after it leaves the prism shown in the diagram below. Make sure to mark all relevant angles.

The angle of 47·7° is greater than the critical angle of 37·6° so the light is internally reflected, and follows the law of reflection.

All the internal angles can be calculated from two pieces of information:

- The angles in a triangle add up to 180°.
- The normal line forms an angle of 90° with the surface of the prism.

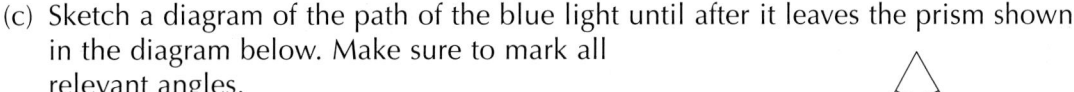

Quick Test 36

1. Define the term 'critical angle'.

2. What happens when the angle of incidence in a material is greater than the critical angle?

Inverse square law

Irradiance

Irradiance is the power of radiation per unit area. It can be expressed in the following relationship:

$$I = \frac{P}{A}$$

where I is the irradiance in watts per metre squared (W m^{-2})

P is the power in watts (W)

A is the area in metres squared (m^2)

Irradiance and distance

Irradiance can be measured with varying distance from a point source. When done, it is discovered that the measured irradiance decreases as the distance from the point source increases. In fact, a graph of irradiance against distance shows an exponential decrease in

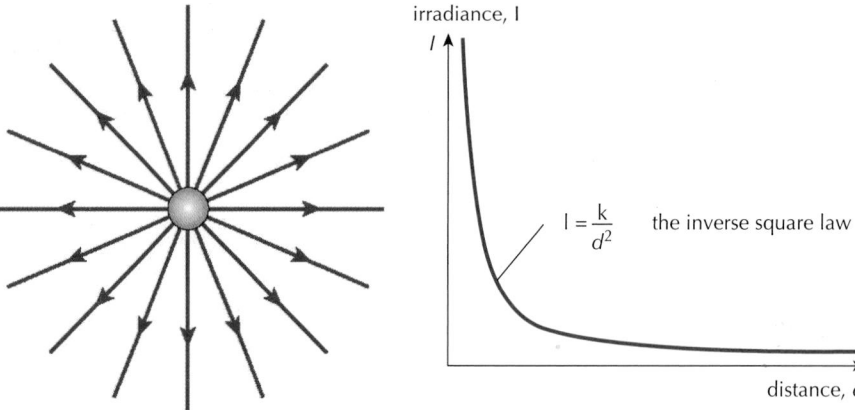

irradiance, I

$I = \dfrac{k}{d^2}$ the inverse square law

distance, d

irradiance with distance as shown below.

This is consistent with the irradiance of radiation from the Sun. The planets closer to the sun, Mercury and Venus for example, experience a greater power per unit area of radiation than the outer planets such as Neptune and Uranus.

A graph of irradiance against $\frac{1}{d^2}$ gives a straight line passing through the origin.

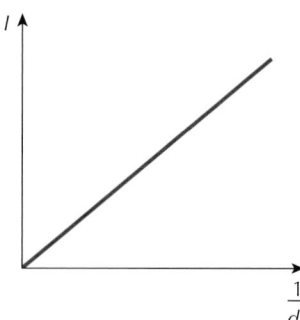

I

$\dfrac{1}{d^2}$

Irradiance and the inverse square law

From the graph above:

$$I \propto \frac{1}{d^2}$$

$$I \propto \frac{k}{d^2}$$

$Id^2 = constant\ (k)$

> **TOP TIP**
>
> This is the equation given in the Higher relationships sheet. You need to be able to derive the final equation from this expression.

This extends to:

$$I_1 d_1^2 = I_2 d_2^2$$

This expression is an inverse square law, meaning that the irradiance varies with the square of the distance from the source, i.e., doubling the distance quarters the irradiance, trebling the distance means the irradiance is 1/9th, and so on.

There are many inverse square laws in physics. They can be used to describe electric fields and gravitational fields among others.

Worked example

A bulb shines light on a screen of area 7·5 m² which is 8·0 m away. The irradiance at the screen is $5W\ m^{-2}$.

(a) Calculate the power incident on the screen

$$I = \frac{P}{A}$$

$$5 = \frac{P}{7 \cdot 5}$$

$$\underline{P = 37 \cdot 5\ W}$$

(b) The screen is moved to 2 m from the bulb. What is the new irradiance on the screen?

$$I_1 d_1^2 = I_2 d_2^2$$

$$5 \times (8)^2 = I_2 \times (2)^2$$

$$320 = 4 I_2$$

$$\underline{I_2 = 80\ W\ m^{-2}}$$

The distance has decreased by a factor of 4, and the irradiance has increased by a factor of 16, as predicted by the inverse square law.

Quick Test 37

1. Define irradiance.
2. Describe how an inverse square law affects irradiance and distance.

Spectra

Emission spectra

Line spectra are emitted when an electric current passes through a low-pressure gas. The electrons collide with atoms in the gas and light is given off. Each element has its own unique line spectrum that can be used to identify it; this acts like an atomic fingerprint.

TOP TIP

A similar effect causes a hot gas to emit light. The line spectra coming from distant stars can be analysed to determine what elements they are made of.

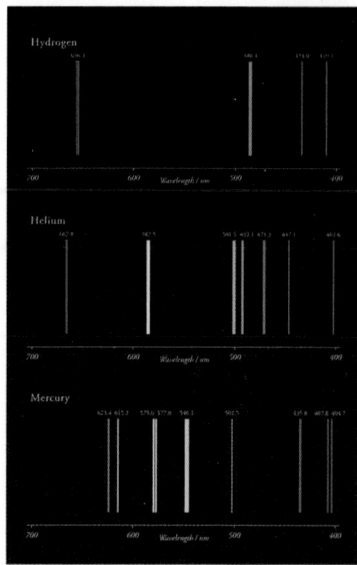

A line emission spectrum can be observed with a spectrometer using a diffraction grating or a prism.

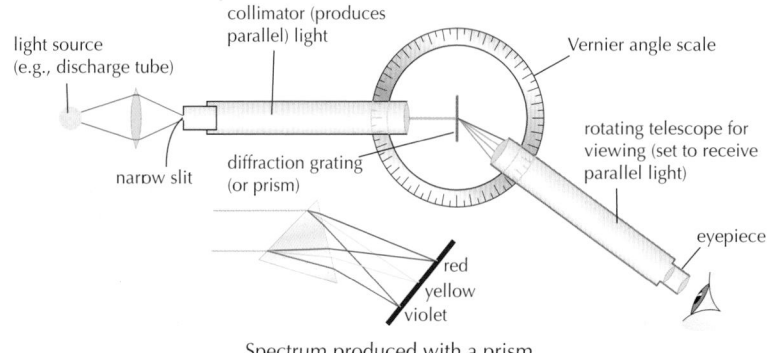

light source (e.g., discharge tube)

collimator (produces parallel) light

Vernier angle scale

rotating telescope for viewing (set to receive parallel light)

narrow slit

diffraction grating (or prism)

eyepiece

red
yellow
violet

Spectrum produced with a prism

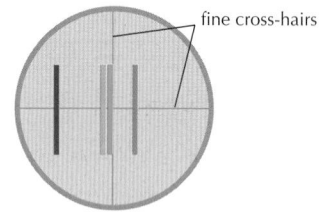

fine cross-hairs

Line spectrum with a grating seen in eyepiece; consists of coloured images of the narrow slit

Each line corresponds to a definite frequency (and so also a wavelength) of radiation.

Absorption spectra

The Sun's mixture of many heated elements appears to produce a continuous spectrum - a spectrum with no visible separate lines. Early in the 19th century, Josef von Fraunhofer discovered that the spectrum of the Sun was crossed with a large number of dark lines.

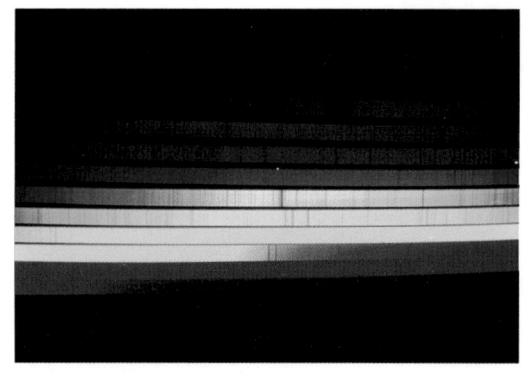

It was shown that the wavelengths associated with these lines corresponded to the emission spectra of known elements. Light emitted from hotter regions of the Sun produces a continuous spectrum, but this light travels through cooler regions in the upper atmosphere. As it does so, atoms in these cooler regions absorb energy from the light at specific wavelengths to produce the absorption lines.

Fraunhofer had discovered a method of determining the elements present in the Sun's atmosphere. We now use similar techniques to identify the elements contained in distant stars and galaxies. Spectra are also used to measure the redshift of galaxies, which allows us to determine the rate of the expansion of the universe.

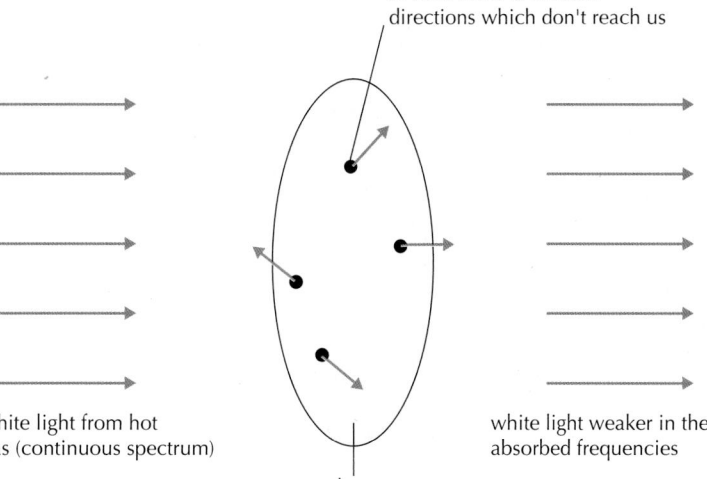

atoms in cooler gas absorb particular frequencies, then re-emit them in random directions which don't reach us

white light from hot gas (continuous spectrum)

white light weaker in the absorbed frequencies

cooler gas

TOP TIP

The line emission spectrum for a particular element has lines corresponding to the same frequency and wavelength of the lines in the absorption spectrum.

Energy levels

As with the photoelectric effect (see page 62), line spectra could not be explained by the wave theory of light. Niels Bohr proposed a revolutionary new model of the atom to explain the production of line spectra. In this new model, electrons are restricted to specific, allowed orbits. Each of these orbits has a certain amount of energy associated with it, so they are often referred to as energy levels. Line spectra can be explained by the transition of electrons between these energy levels. Because the electrons in the atoms of each element are arranged differently, electron transitions produce a unique line spectrum for each element.

Quick Test 38

1. State what is meant by a continuous spectrum.

2. Explain how absorption lines in the spectrum of sunlight allow us to determine the elements present in the Sun's atmosphere.

The model of the atom

The Bohr model

In 1913, Danish physicist, Niels Bohr, proposed a new model of the atom. In Bohr's model, electrons orbit at different distances from the nucleus, corresponding to their energy levels. There is a maximum number of electrons for each energy level. Electrons tend to occupy the lowest energy level, i.e., the orbits closest to the nucleus. Electrons can move between the orbits, but cannot stop in between. The different orbits can be represented in an energy level diagram as shown to the right.

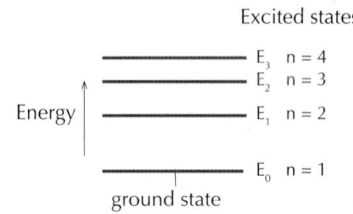

An electron in the lowest energy level is said to be in the **ground state**. An electron that moves to a higher energy level is said to be in an **excited state**. If the electron gains sufficient energy, it can escape from the atom completely and reach the **ionisation level**.

The electron is said to have zero energy when it has reached the ionisation level. The energy levels in the atom therefore have negative energy. When an electron gains energy and moves to a higher energy level, it has less negative energy. It is closer to having zero energy.

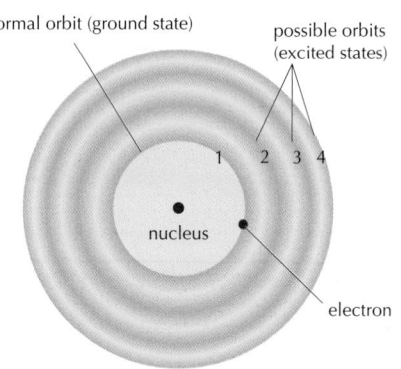

The Bohr model

TOP TIP

The energy will be shown in energy level diagrams as a negative number. Remember this when calculating the difference in energy (ΔE) between levels.

Electron transitions

Electron transitions occur between the energy levels of an atom when an electron either absorbs or emits a photon of electromagnetic radiation. An electron can only move from one energy level to another by gaining or losing **exactly** the right amount of energy. The energy absorbed by the electron must match the gap between energy levels.

Line emission spectra are produced when electrons make the transition from a higher energy level to a lower energy level. When the electron moves to a lower level it loses energy. This energy is released in the form of a photon, which may or may not lie in the visible spectrum.

The frequency of the emitted photon is related to its energy by the relationship:

$$E = hf$$

where h is Planck's constant

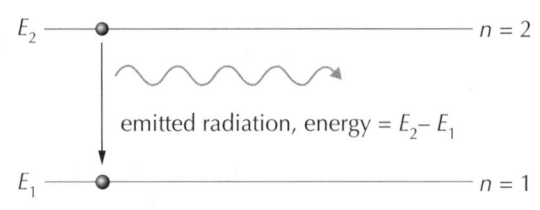

In the diagram shown here:

$$E_2 - E_1 = \Delta E = hf$$

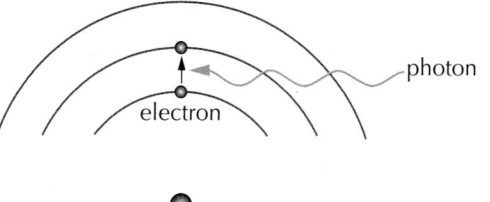

photon   The electron
to a higher-e
by absorbing a photon
of exactly the right
energy.

electron

nucleus

Line absorption spectra are produced when electrons make the transition from a lower energy level to a higher energy level. The electron is excited to a higher energy level by absorbing a photon of **exactly** the right energy.

An excited electron will eventually return to the ground state. The electron can jump between more than one energy level at a time. The bigger the jump between levels, the higher the frequency of the photon emitted.

> **Top Tip**
>
> Energy is directly proportional to frequency.

Worked example

The energy level diagram of an element is shown here.

(a) Which transition between these energy levels will produce radiation with the shortest wavelength?

E_3 ——————— -3.2×10^{-19} J

E_2 ——————— -1.6×10^{-19} J
E_1 ——————— -0.8×10^{-19} J

E_0 ——————— -3.2×10^{-18} J

E_3 to E_0

There is the greatest energy difference between these levels. Since E = hf, the largest energy difference will produce radiation of the highest frequency and therefore the shortest wavelength.

(b) An electron makes the transition from energy level E_2 to E_3 by absorbing light energy. What frequency of light is used to cause this transition?

$$\Delta E = E_3 - E_2$$
$$\Delta E = 3\cdot2 \times 10^{-19} - 1\cdot6 \times 10^{-19}$$
$$\Delta E = 1\cdot6 \times 10^{-19} J$$

$$E = hf$$
$$1\cdot6 \times 10^{-19} = 6\cdot63 \times 10^{-34} f$$
$$\underline{f = 2.4 \times 10^{14} Hz}$$

Quick Test 39

1. State what is meant by the 'ground state' in the Bohr model of the atom.
2. State what is meant by the 'ionisation level' in the Bohr model of the atom.
3. Describe how an electron can make the transition to a higher energy level.
4. Describe the process of an electron moving from a higher energy level to a lower energy level.

Alternating current

Measuring alternating current

An alternating current (AC) periodically changes direction. This is in contrast to direct current (DC), which only travels in one direction. We can represent alternating current as a **sinusoidal waveform**. This shows that the instantaneous value of voltage changes with time. Examples of AC and DC can be examined using an oscilloscope:

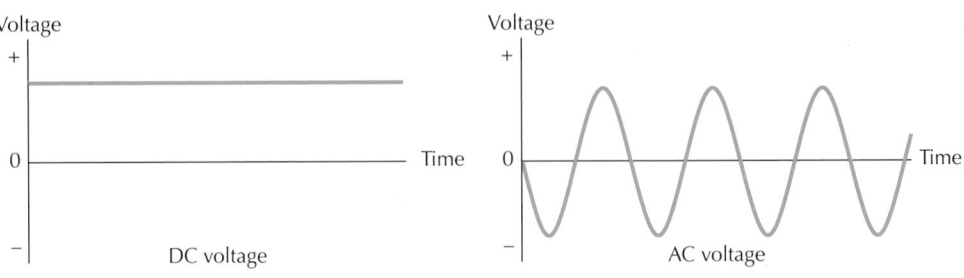

DC voltage

AC voltage

Measuring frequency and peak voltage

An oscilloscope displays how the voltage changes over time. In a similar way to a graph, it shows voltage on the y-axis and time on the x-axis. The difference between a graph and an oscilloscope is that instead of having a scale along each axis, the oscilloscope controls the values of voltage and time via dials adjacent to the screen. A dial called 'volts/div' controls the voltage axis, and a dial called the 'timebase' controls the time axis.

The frequency of an AC wave can be determined from its period, which is the time for one wave. The period can be calculated if the setting on the timebase dial is known.

If the setting on the volts/div dial is known we can determine the peak voltage. The peak voltage corresponds to half the vertical height of the wave.

Example

An AC supply is connected to an oscilloscope, as shown here.

Calculate:

(a) The peak voltage

The volts/div dial indicates 2 volts/div. The wave height is 6 divisions in total, so the peak voltage corresponds to 3 divisions.

$$V_{peak} = 3 \times 2 = 6\,V$$

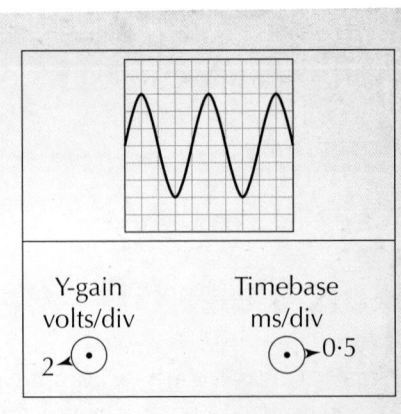

(b) The frequency of the AC supply

The timebase dial indicates 0·5 ms/div. It takes 4 divisions to complete one cycle. The period T is therefore:

$$T = 4 \times 0 \cdot 5 \times 10^{-3} = 2 \times 10^{-3} s$$

$$f = \frac{1}{T} = \frac{1}{2 \times 10^{-3}} = 500 \, Hz$$

TOP TIP

Be careful of the units used for timebase settings, i.e., milliseconds (ms). A common mistake is to overlook the prefix and incorrectly use seconds.

Peak and RMS

The value of the mains supply in the UK is 230 V. But what does this actually mean? From a sinusoidal AC waveform, it can be seen that the instantaneous voltage at certain points is zero, a peak value, or some value in between. The value of 230 V is the effective value of the mains supply, and is called the root mean square (RMS).

The definition of the RMS voltage is the value of direct voltage that produces the same power (e.g., heating or lighting), as the alternating voltage. For example, a 12 V battery will produce the same brightness in a lamp as a 12 V RMS AC supply.

The relationship between the peak voltage and the RMS voltage is:

$$V_{peak} = \sqrt{2} \, V_{rms}$$

The same relationship holds for peak and RMS current:

$$I_{peak} = \sqrt{2} \, I_{rms}$$

TOP TIP

V_{peak} is always greater than V_{rms}

Quick Test 40

1. Describe how to calculate the frequency of an AC waveform from the trace on an oscilloscope.

2. The peak value of an AC voltage is 12 V. Calculate the RMS voltage.

3. Research the advantages and disadvantages of AC and DC.

4. The timebase setting on an oscilloscope is 0·5 ms/div. There are 4 waves displayed on the screen. The timebase setting is changed to 1·0 ms/div. Describe the effect on the number of waves displayed on the screen.

Circuits revision

Basic definitions

Electric current is the flow of charge (normally electrons). The current is the total charge that passes a point in a circuit per second. $I = \dfrac{Q}{t}$

where I is the current in amps (A) Q is the charge in coulombs (C) t is the time in seconds (s)

Potential difference (V) is the energy supplied to each coulomb of charge.
Potential difference is measured in joules per coulomb ($J\ C^{-1}$) or volts.

$$1 \text{ joule per coulomb} = 1 \text{ volt}$$

Power is the rate at which energy is transferred per second.

$$P = \frac{E}{t}$$

where P is the power in watts (W) E is the energy in joules (J) t is the time in seconds (s)

Power can also be calculated from $P = IV$, $P = I^2R$ and $P = \dfrac{V^2}{R}$

Resistance is the opposition to electric current and is measured in ohms (Ω).
Resistance is defined by the equation: $R = \dfrac{V}{I}$

Series circuit rules

Current:	Voltage:	Resistance:
$I_s = I_1 = I_2 = I_3$	$V_s = V_1 + V_2 + \cdots$	$R_T = R_1 + R_2 + \cdots$

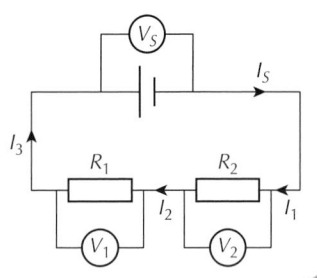

Parallel circuit rules

Current:	Voltage:	Resistance:
$I_s = I_1 + I_2 + \cdots$	$V_s = V_1 = V_2 = \cdots$	$\dfrac{1}{R_T} = \dfrac{1}{R_1} + \dfrac{1}{R_2} + \cdots$

TOP TIP

Only the circuit rules for resistance are given in the Higher relationships sheet. The rules for current and voltage must be remembered.

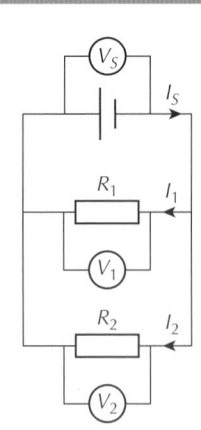

Worked example

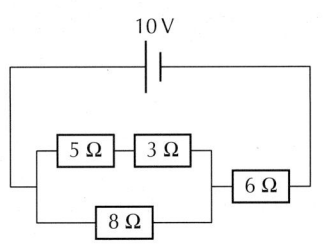

(a) Calculate the current in each resistor.

First we must calculate the total resistance of the circuit. The 5Ω and 3Ω resistors are in series and so are equivalent to a single 8Ω resistor.

Calculate the resistance of the parallel branch:

$$\frac{1}{R_p} = \frac{1}{R_1} + \frac{1}{R_2}$$

$$\frac{1}{R_p} = \frac{1}{8} + \frac{1}{8}$$

$$R_p = 4\Omega$$

Then calculate the total resistance:

$$R_T = 4 + 6 = 10\ \Omega$$

$$I = \frac{V_S}{R_T} = \frac{10}{10} = 1\ A$$

Current in 6Ω resistor = 1 A

Both parallel branches have the same resistance and so experience the same value of current:

Current in 8Ω resistor = 0·5 A

Current in 3Ω and 5Ω resistor = 0·5 A

(b) *Calculate the potential difference across each resistor*

$$6\Omega : V = IR = 1 \times 6 = 6V$$

$$8\Omega : V = IR = 0 \cdot 5 \times 8 = 4V$$

$$5\Omega : V = IR = 0 \cdot 5 \times 5 = 2 \cdot 5V$$

$$3\Omega : V = IR = 0 \cdot 5 \times 3 = 1 \cdot 5V$$

(c) Calculate the power dissipated in each resistor

$$6\Omega : P = IV = 1 \times 6 = 6W$$

$$8\Omega : P = \frac{V^2}{R} = \frac{4^2}{8} = 2W$$

$$5\Omega : P = I^2 R = 0 \cdot 5^2 \times 5 = 1 \cdot 25W$$

$$3\Omega : P = IV = 0.5 \times 1 \cdot 5 = 0 \cdot 75W$$

Quick Test 41

1. Describe the rules for voltage in series and parallel circuits.
2. Describe the rules for current in series and parallel circuits.
3. List the equations that can be used to calculate power in circuits.

The Wheatstone bridge

Potential difference

Any power supply (electron) current leaves one terminal with a low potential and returns to the other at a higher potential. The individual potentials of the terminals are not important - only the potential difference determines the circuit current.

For example:

$$9\,V \quad\quad -3\,V$$

The potential difference across the cell is 12V. The same potential difference could be achieved if the individual potentials were 10V and −2V or 18V and 6V.

Potential dividers

A potential divider allows access to a fixed voltage which is smaller than the supply voltage used by the circuit, e.g., obtaining 3V from a 9V supply.

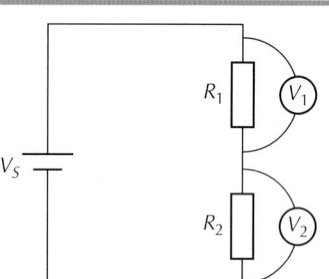

For a potential divider circuit:

$$\frac{V_1}{V_2} = \frac{R_1}{R_2}$$

The voltage across R_1 is given by:

$$V_1 = \left(\frac{R_1}{R_1 + R_2}\right) V_S$$

The voltage across R_2 is given by:

$$V_2 = \left(\frac{R_2}{R_1 + R_2}\right) V_S$$

Consider both the circuits below:

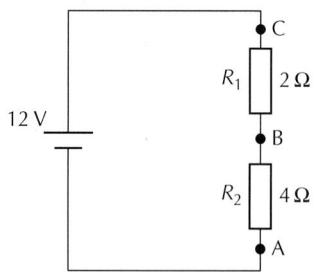

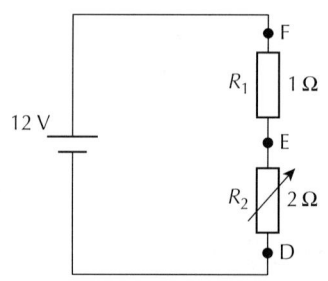

Using the equations on the previous page, we can determine the voltages across each of the resistors:

V_{BC} = 4V $\qquad\qquad$ V_{EF} = 4V

V_{AB} = 8V $\qquad\qquad$ V_{DE} = 8V

so $\quad$ potential at A = 0V $\qquad$ so $\quad$ potential at D = 0V

$\qquad$ potential at B = 8V $\qquad\qquad$ potential at E = 8V

$\qquad$ potential at C = 12V $\qquad\qquad$ potential at F = 12V

The balanced Wheatstone bridge

If we combine the two previous potential divider circuits and place a voltmeter between points B and E, we form a Wheatstone bridge.

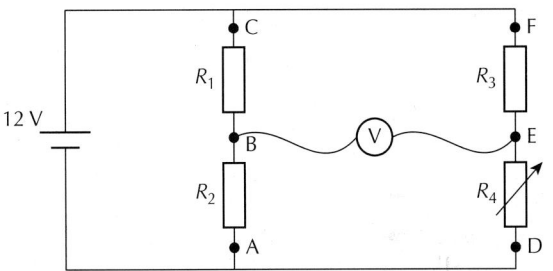

Since the potential at points B and E is 8V, the potential difference between B and E is 0V. When the reading on the voltmeter is zero, the bridge is said to be balanced.

For a balanced Wheatstone bridge:

$$\frac{R_1}{R_2} = \frac{R_3}{R_4}$$

If the reading on the voltmeter is non-zero then the bridge is said to be out of balance.

TOP TIP

Care must be taken in identifying the correct values of R_1, R_2, R_3 and R_4 in a Wheatstone bridge. It helps to think of the Wheatstone bridge as two separate potential dividers where R_1 and R_2 make up one pair or resistors and R_3 and R_4 make up the other pair.

Quick Test 42

1. What is meant by saying that a Wheatstone bridge is balanced?

2. What changes could be made to a balanced Wheatstone bridge circuit to make the bridge out of balance?

Sources of electricity

Electrical sources

In most situations, we assume batteries and power supplies to be ideal. An ideal power supply is one in which the voltage remains constant whether current is drawn or not. By choosing appropriate values of resistors we can produce whatever value of current we like. These assumptions work reasonably well for the majority of cases, but in order to make our calculations truly accurate we must also take into account the resistance of the power supply itself.

The materials inside a battery or power supply offer a resistance to the passage of current. This is known as the internal resistance of the supply, and is due to collisions between charged particles and other atoms within the power supply. In most cases the internal resistance is so small that it can be considered negligible. However, when the current is large the effects of internal resistance can be significant.

Electromotive force

The voltage of an ideal power supply is called the electromotive force (EMF). The EMF is defined as the potential difference across the source when no current is drawn (open circuit), and is the energy given to each coulomb of charge as it passes through the power supply. For example, a cell of EMF 1·5 V supplies 1·5 J of energy to each coulomb of charge that passes through the cell.

We can now represent a battery, or cell as:

E = EMF

r = internal resistance

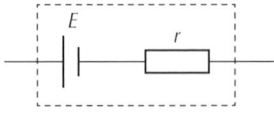

TOP TIP

Inside the dotted line can be thought of as the 'internal' part of the circuit. Anything outside of that dotted line is the 'external' circuit.

Terminal potential difference and 'lost' volts

The effect of the internal resistance is to reduce the potential difference across the external circuit. When a very high resistance voltmeter is connected directly to a cell, the reading on the voltmeter is the cell's EMF as no current in flowing.

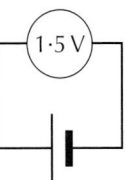

When the cell is connected to a circuit, the voltmeter will read a value that is less than 1·5 V. The reading on the voltmeter is called the terminal potential difference (TPD).

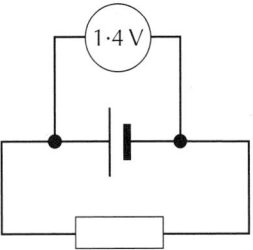

The voltage drop is caused by the internal resistance of the cell and is called the 'lost' volts. The 'lost' volts in this case is 0·1 V. Therefore:

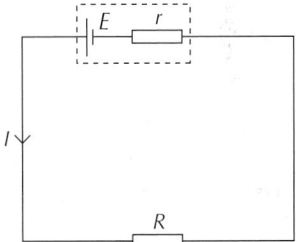

$$\text{EMF} = \text{TPD} + \text{'lost' volts}$$
$$E = V + Ir$$
$$E = IR + Ir$$
$$E = I(R + r)$$

Quick Test 43

1. State what is meant by the electromotive force.
2. State what is meant by the terminal potential difference.
3. Explain why volts are 'lost' when current is drawn from a power supply by an external circuit.

Internal resistance

Graphical analysis of EMF and internal resistance

The circuit shown can be used to determine the EMF and the internal resistance of the cell. The variable resistor is adjusted to enable a range of readings for current and potential difference. A graph drawn of the results with V on the y-axis and I on the x-axis gives a straight line of negative gradient. This is consistent with the formula:

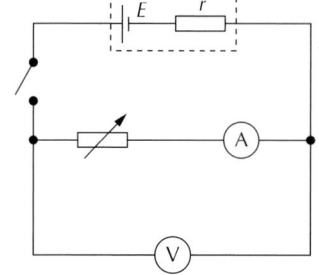

$$V = E - Ir$$

$$V = -rI + E$$

Compare to the equation of a straight line:

$$y = mx + c$$

V = y-axis $E = c$

I = x-axis $-r = m$

We can see that gradient = –internal resistance and EMF = y-intercept

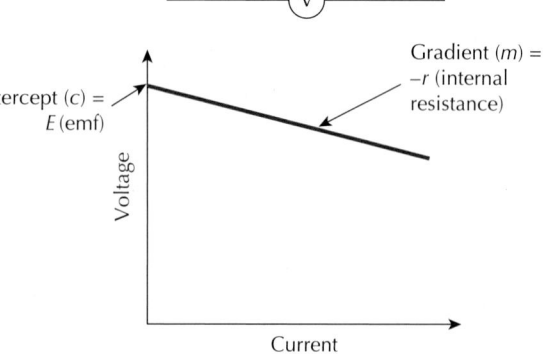

Intercept (c) = E (emf)

Gradient (m) = $-r$ (internal resistance)

Voltage

Current

TOP TIP

The x-intercept on a graph of V vs I will give the value of the short circuit current.

Short circuit current

The maximum current available from a power supply is the short circuit current. This is the current that is present when the terminals of the power supply are joined by a short piece of wire.

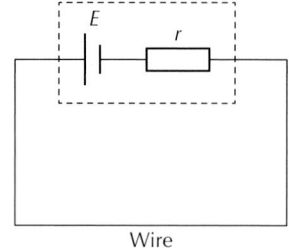

Wire

In this case there is no external resistance (assuming the resistance of the short wire is negligible) so R = 0 Ω and therefore V = IR = 0 V (no TPD).

By substituting R = 0 Ω into the internal resistance equation we find:

$$E = I(0 + r)$$

$$E = Ir$$

$$I_{max} = \frac{E}{r}$$

Worked example

A circuit contains a battery of EMF 4 V and internal resistance 1·5 Ω. The external resistance is 6·5Ω.

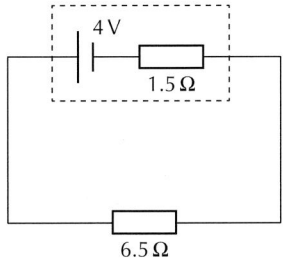

4 V
1.5 Ω
6.5 Ω

(a) Calculate the current drawn from the power supply.

$E = I(R+r)$

$4 = I(6\cdot5+1\cdot5)$

$4 = 8I$

$\underline{I = 0\cdot5\ A}$

(b) Calculate the terminal potential difference.

$V = IR$

$V = 0\cdot5\times6\cdot5$

$\underline{V = 3\cdot25\ V}$

(c) Calculate the short circuit current.

$I_{max} = \dfrac{E}{r}$

$I_{max} = \dfrac{4}{1\cdot5}$

$\underline{I_{max} = 2\cdot67\ A}$

Quick Test 44

1. Describe how to obtain the internal resistance and EMF from a graph of V vs I.
2. Explain how a maximum current can be achieved in a circuit

Defining capacitance

Capacitors

A capacitor is a component which stores charge.
It can be identified by the circuit symbol:

A capacitor is composed of layers of conducting
material which are separated by an insulator
- two metal plates separated by an air gap,
for example. Capacitors come in all shapes
and sizes, with the majority of them being
cylindrical.

Relationship between charge and potential difference

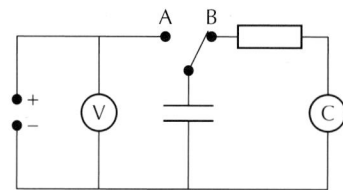

When the switch is in position A, the power supply charges the capacitor to a chosen
voltage. When the switch is moved to position B, the capacitor discharges through the
resistor and coulombmeter. We can then obtain pairs of values for charge (Q) at a given
voltage (V).

The graph of Q against V gives a straight line passing
through the origin, indicating direct proportion.

$$Q \propto V$$

$$Q = kV$$

$$k = \text{constant} = \text{capacitance (C)}$$

$$Q = CV$$

Therefore $C = \dfrac{Q}{V}$ where the units of capacitance are farads (F).

$$1 \text{ F} = 1 \text{ C V}^{-1}$$

The definition of capacitance is therefore the charge
stored per volt. This means that capacitance is the
gradient of the line in a Q against V graph.

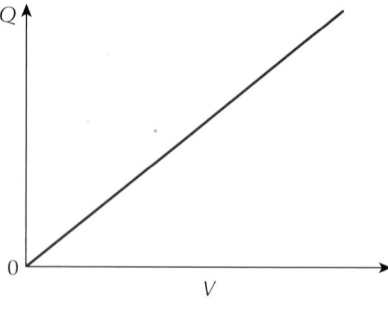

TOP TIP

1 farad is too large for practical
use. Realistic capacitor sizes are
microfarads (μF), nanofarads (nF)
and picofarads (pF).

Worked example

A capacitor stores 1×10^{-3} C of charge when the potential difference across it is 100 V. Calculate the capacitance.

$$C = \frac{Q}{V}$$

$$C = \frac{1 \times 10^{-3}}{100}$$

$$C = 1 \times 10^{-5} F$$

This is a value of 10 μF.

Quick Test 45

1. State the definition of capacitance.
2. What is the relationship between capacitance, charge and potential difference?
3. How can capacitance be calculated from a Q against V graph?

Energy storage

Electron flow in a capacitor circuit

- Initially the capacitor is uncharged.
- Electrons pass from the negative terminal of the battery to one plate of the capacitor and away from the other plate.
- Voltage across the capacitor (V_c) is initially zero, and then increases.

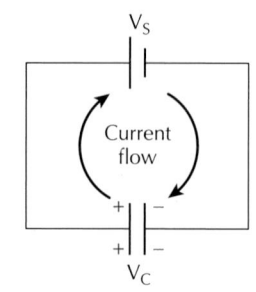

- More negative charge is on one plate of the capacitor. The increased charge repels more negative charges arriving, hence the current drawn becomes less. The increasing positive charge on the other plate of the capacitor makes the removal of electrons from that plate even more difficult as time passes, due to electrostatic attraction. Current also decreases as the negative charge on the plate repels the arrival of more electrons being "pushed" by the battery.
- Potential difference across the capacitor increases as the build-up of charge on both plates increases.
- Eventually current decreases until none passes. This is because the potential difference across the plates of the capacitor is equal to the potential difference across the battery
  ($V_s = V_c$)

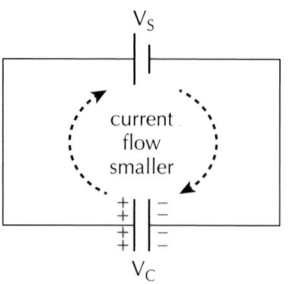

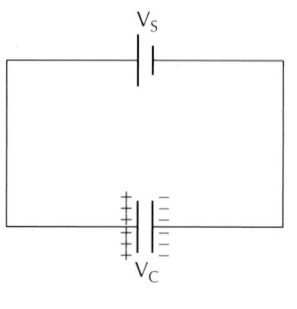

TOP TIP

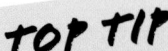

Current does not pass through the capacitor as there is an insulating gap. Electrons pass onto one plate and away from the other.

Energy stored in a capacitor

The negatively charged plate repels the electrons approaching it and the positively charged plate attracts the electrons moving away from it. In order to overcome the forces of attraction and repulsion, work has to be done and energy has to be supplied. This energy is supplied by the battery and is stored by the capacitor.

The energy stored by a capacitor can be used to maintain the memory in a calculator, or provide a constant current in a circuit where the supply voltage is too low.

The amount of energy stored by a capacitor depends on the amount of charge on the plates of the capacitor and the capacitance of the capacitor. The graph below shows how the voltage across the plates of a capacitor changes with the charge stored.

The shaded area under the graph shown represents the total energy stored in the capacitor.

$$Area\ under\ graph = \frac{1}{2}QV$$

$$E = \frac{1}{2}QV$$

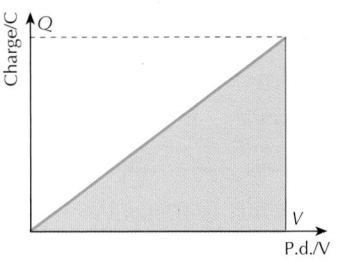

where E is the energy stored in joules (J)

Q is the charge in coulombs (C)

V is the potential difference in volts (V)

TOP TIP

Be careful not to confuse $E = \frac{1}{2}QV$ with $W = QV$ which only applies to a **fixed** potential difference between two points and is never used with capacitors.

Energy formula

Since $Q = CV$ there are two other versions of the energy formula for a capacitor.

$$E = \frac{1}{2}QV \ \text{but} \ Q = CV$$

$$\therefore \ \ E = \frac{1}{2}(CV)V$$

$$E = \frac{1}{2}CV^2$$

$$E = \frac{1}{2}QV \ \text{but} \ V = \frac{Q}{C}$$

$$\therefore \ \ E = \frac{1}{2}Q\left(\frac{Q}{C}\right)$$

$$E = \frac{1}{2}\frac{Q^2}{C}$$

Quick Test 46

1. Why does energy have to be supplied in order to charge a capacitor?

2. How can the total energy stored in a capacitor be calculated from a charge against voltage graph?

Capacitors in DC circuits

Charging a capacitor

When the switch is closed in the circuit shown, charge begins to flow to one plate of the capacitor from the other, via the battery/cell in an anti-clockwise direction. As charge builds up on the plates of the capacitor, the potential difference across the plates increases as the current gradually decreases. When the capacitor is fully charged the potential difference across the plates of the capacitor is equal to the supply voltage and the current reduces to zero. These effects are shown in the following graphs:

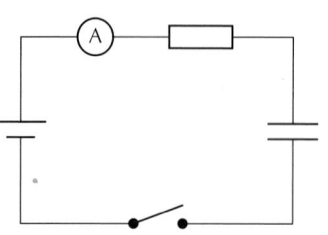

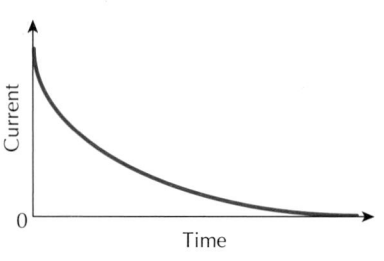

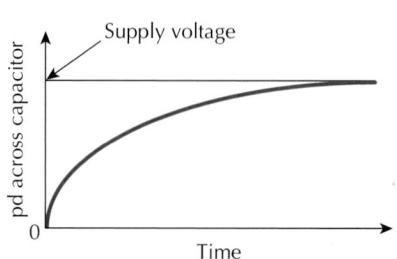

TOP TIP

You need to remember the correct current against time and potential difference against time graphs for both charging and discharging capacitors.

Discharging a capacitor

As with the charging process, the current starts out at a maximum value before decreasing to zero as the capacitor discharges through the resistor. However, the passage of current is in the opposite direction as no charges can flow through the insulator between the plates of the capacitor.

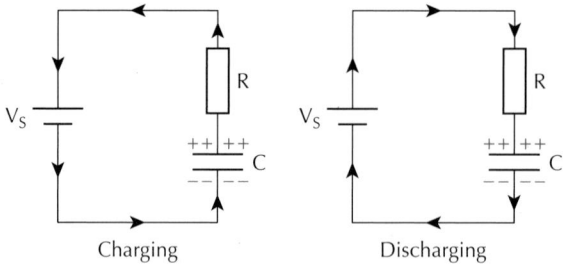

As the current decreases, the potential difference across the plates of the capacitor decreases as well. This is because there is less charge on the plates of the capacitor. When the capacitor is fully discharged the current and the potential difference across the capacitor are both zero.

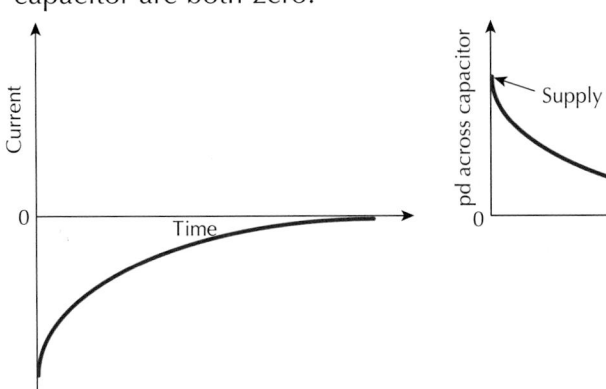

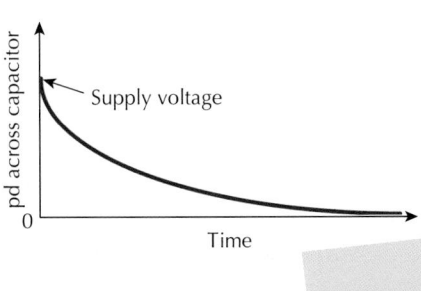

TOP TIP

The maximum current is determined by the value of the series resistor and can be calculated using Ohm's law (V=IR).

Factors affecting the rate of charge and discharge

The time taken for a capacitor to charge is determined by the values of resistance and capacitance. Larger values of capacitance and larger values of resistance will increase the time it takes a capacitor to charge or discharge.

Current–time graphs for differing values of capacitance and resistance are shown below for a charging capacitor:

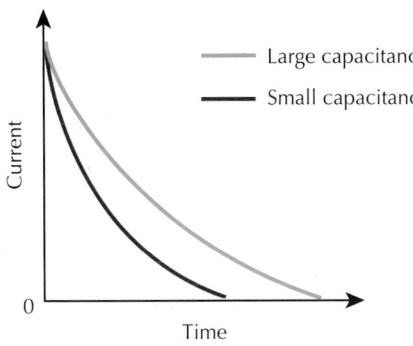

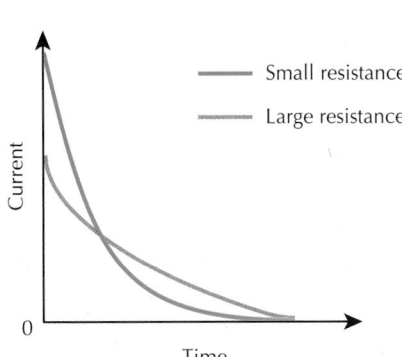

Note that the initial current (charging current) is smaller for a larger value of resistance.

RC circuits

The switch S in the following circuit is closed.

Immediately after the switch is closed:

- The charge on the capacitor is 0 C.
- The potential difference across the plates of the capacitor is 0 V.
- The potential difference across the resistor is 12 V.
- The charging current is:
$$V = IR$$
$$12 = I \times 1 \times 10^6$$
$$I = 1 \cdot 2 \times 10^{-5} \ A$$

When the capacitor is fully charged:

- The potential difference across the plates of the capacitor is 12 V.
- The charge stored by the capacitor is:
$$Q = CV$$
$$Q = 2200 \times 10^{-6} \times 12$$
$$Q = 0 \cdot 026 \ C$$

When the potential difference across R is 4 V:

- The potential difference across the plates of the capacitor is 8 V since VS = VC + VR at any time during charging.
- The charge stored by the capacitor is:
$$Q = CV$$
$$Q = 2200 \times 10^{-6} \times 8$$
$$Q = 0 \cdot 018 \ C$$

TOP TIP

An RC circuit has both a resistor and capacitator.

Quick Test 47

1. Describe the effect of changing capacitance on the time it takes to charge and discharge a capacitor.

2. Describe the effects of increasing resistance on the maximum value of current in an RC circuit.

Practical applications

Flash photography

Capacitors are used in flash cameras to briefly power a high voltage flash tube. Flash tubes require very high currents for a very short period of time. A capacitor can be discharged through a flash tube, thus providing a short pulse of electrical energy that does not generate a large amount of heat as the current quickly decreases.

Smoothing

Capacitors can be used to smooth or even out fluctuations in an electrical signal. DC power supplies often produce voltages that fluctuate greatly. When a steady DC signal is needed, a smoothing capacitor can be used to smooth out the fluctuating signal.

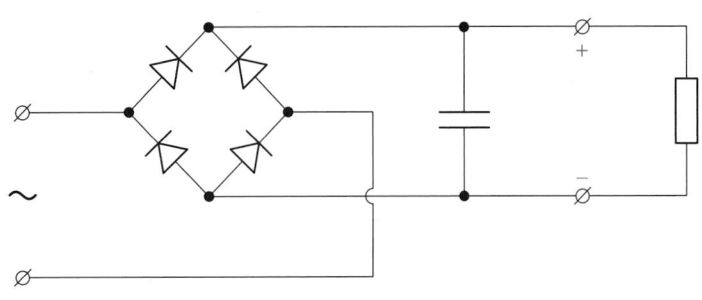

A smoothing capacitor can be used with a diode bridge shown here to convert alternating current to direct current.

Noise suppression

When sparks occur, for example, at the contacts of switches, high frequency electromagnetic waves are produced. These electromagnetic waves can cause interference with other electromagnetic radiation, such as radio and television signals. By fitting a capacitor across the contacts of switches, much of the high frequency pulse can be safely bypassed, reducing the interference produced.

Capacitance based touch screens

Some tablet computers, interactive whiteboards and mobile telephones use a technology called capacitive sensing, to enable touchscreens to be used as input devices. A layer that stores electrical charge is placed on the screen. When the screen is touched, a small amount of that charge is transferred to the finger thus reducing the overall charge on the screen. The device uses these small changes in capacitance to detect where the event took place on the screen, and the touchscreen software tells the device what do with that information, whether it be to open an app, zoom in on a photo, etc.

Capacitive touchscreens will not operate with a standard stylus or through clothing because the object in contact with the screen must be a conductor that can pick up charge from the screen. You can purchase capacitive gloves that have a conducting layer on the fingertips to keep your hands warm while you use your smart devices!

Quick Test 48

1. What makes a capacitor a useful device in flash photography?
2. List another practical use of capacitors.

Classifying materials

Conductors, insulators and semiconductors

All materials can be classified into three groups: conductors, insulators and semiconductors. Each of the three groups has different electrical properties.

Conductors	Materials with many free electrons. These electrons can flow easily through the material. e.g., metals, graphite
Insulators	Materials with very few free electrons. e.g., plastics, wood, glass
Semiconductors	Materials which lie somewhere between conductors and insulators in terms of their conductivity. They are insulators when in their purest form, but will conduct when an impurity is added. e.g., silicon, germanium, gallium arsenide

Band theory

The electrons in an atom occupy specific energy levels. When atoms join to form solids they can use the energy levels of neighbouring atoms. When a large number of atoms join together, their many different energy levels become grouped together in **energy bands**. There are spaces between these energy bands called **energy gaps**.

Electrons will occupy the lower energy levels closest to the nucleus first. In order to move to a higher energy level an electron must obtain energy. In terms of electrical conductivity there are two energy bands that are of particular importance. These bands are called the **valence band** and the **conduction band**. The valence band is the highest energy band that electrons will normally occupy at room temperature. The conduction band is the highest occupied energy band above the valence band. The conduction band can accept electrons from the valence band under the right conditions. Electrons in the conduction band are free to move.

TOP TIP

Energy bands can be thought of in a similar way to the energy levels in the Bohr model of the atom.

Conductor energy bands

In conductors, the valence band and the conduction band overlap, which allows the valence electrons to move freely through the material.

Conduction band
Valence band

Valence and conduction
bands overlap

Insulator energy bands

In insulators, the valence band is full of electrons, but there is a large energy gap between it and the conduction band. At room temperature there is not enough energy to move the electrons from the valence band to the conduction band, and so insulators do not normally conduct electricity.

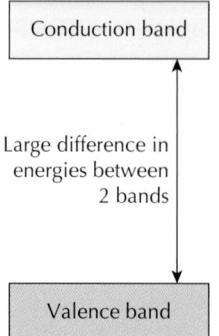

If the temperature is high enough or the supplied voltage is sufficiently large, some electrons can be lifted to the conduction band to allow current to pass, but this will often damage the material.

Semiconductor energy bands

In semiconductors the energy gap between the valence band and the conduction band is relatively small. At room temperature there is sufficient energy to move electrons from the valence band to the conduction band, allowing current to pass. An increase in the temperature of a semiconductor will increase the conductivity of a semiconductor.

Conduction band

Valence band

Small difference in energy between 2 bands

Quick Test 49

1. The electrical properties of materials can be classified into three groups. Name these three groups.

2. State what is meant by the *valence band*.

3. State what is meant by the *conduction band*.

4. Explain why insulators cannot conduct electricity at room temperature but semiconductors can.

Semiconductors

Bonding in semiconductors

Materials used to make semiconductors usually have a valency of 4, like silicon or germanium for example. These materials have four outer electrons available for bonding. Elements of this type bond covalently to form a crystal lattice structure, where each of the electrons is bonded to another atom. This means that there are very few free electrons available to conduct.

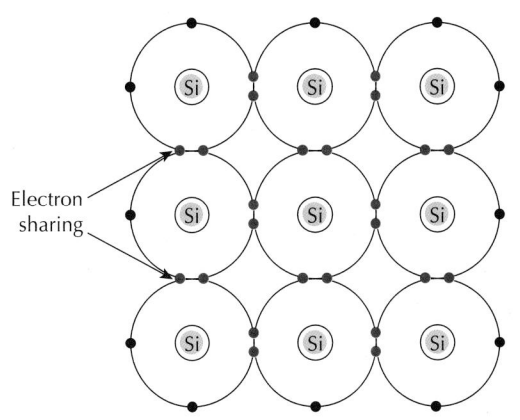

Electron sharing

The resistance of the crystal structure is very high. Increasing the temperature of the semiconductor will result in more free electrons, thus reducing the resistance.

Holes

When an electron moves its position in the crystal lattice it leaves a space behind it that is positively charged. This absence of an electron is called a positive hole. An electron will move to fill any positive holes in the crystal structure, and in turn leave a positive hole. Although it is electrons that are moving in the semiconductor, it is easier to think in terms of the movement of holes. The hole can therefore be thought of as a positive charge carrier.

Doping

The electrical properties of semiconductors can be changed by the addition of very small amounts of impurities. The process of adding impurities to semiconductors is known as doping. Doping can increase the conductivity of a semiconductor in two different ways:

1. n-type semiconductors

One in every million or so silicon atoms is replaced with an atom such as arsenic (As), which has five outer electrons. Four of its outer electrons will bond covalently with the silicon, leaving the fifth free to move and conduct. The conductivity of the crystal lattice structure has increased with the addition of an impurity, therefore decreasing the resistance of the semiconductor.

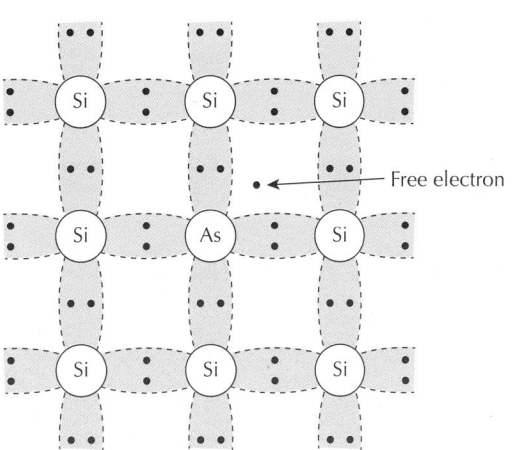

Free electron

This type of semiconductor is called n-type because the majority charge carriers are electrons which are negatively charged.

2. p-type semiconductors

One in every million or so silicon atoms is replaced with an atom such as boron (B), which has three outer electrons. These three outer electrons will bond covalently with the silicon, producing a positive 'hole' where there is an absence of an electron. An electron will move to fill this hole, and conduction can take place through the movement of positive holes. Again, the conductivity of the crystal lattice structure has increased with the addition of an impurity, therefore decreasing the resistance of the semiconductor.

This type of semiconductor is called p-type because the majority charge carriers are holes which are positively charged.

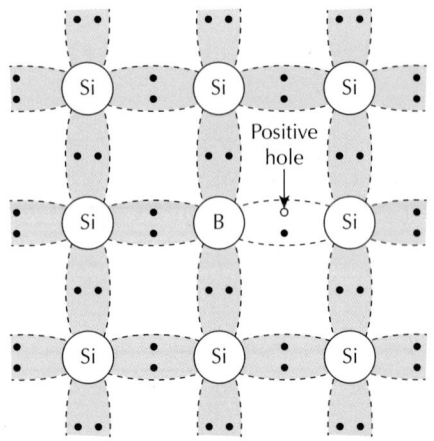

TOP TIP

Although p-type and n-type materials have different charge carriers, they are both still electrically neutral. The addition of doping agents such as arsenic or boron does not affect the overall charge of the semiconductor because in the case of arsenic, although it has an additional electron which it can use as a charge carrier, it also has an additional proton in its nucleus. Similarly, although boron has one less electron its nucleus has one less proton so it is still electrically neutral.

Quick Test 50

1. Explain what is meant by doping.
2. Describe how an n-type semiconductor is created.
3. Describe how p-type semiconductor is created.
4. What effect does doping have on the resistance of the semiconductor?

The p-n junction

Diodes

A semiconductor can be grown so that one half is made of p-type material and the other half is made of n-type material. This is called a p-n junction and is the basis of diodes, which can be used in solar cells, light dependent resistors and thermistors.

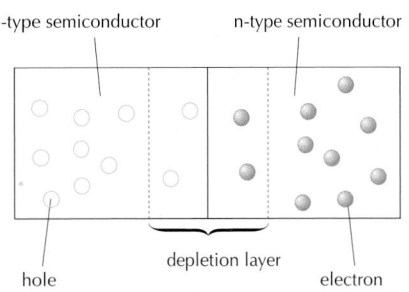

p-type semiconductor n-type semiconductor

depletion layer

hole electron

Very close to the junction of the two materials, the excess holes from the p-type material combine with the excess electrons from the n-type material. The region where the charges combine is called the depletion layer.

This creates a region of positive charge on the side of the n-type material and a region of negative charge on the side of the p-type material, resulting in an area of no charge carriers. Since like charges repel it is now difficult for charges to move across the depletion layer.

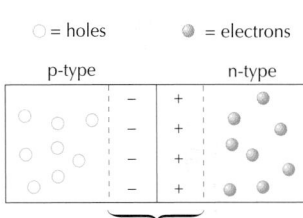

○ = holes ● = electrons

p-type n-type

depletion layer (exaggerated width)

Biasing the diode

When we apply an external voltage to the diode we say that the diode is biased. There are two way to bias the diode:

1. Reverse biased

To reverse bias the diode, a positive voltage is applied to the n-type side of the material and a negative voltage is applied to the p-type side of the material. This has the effect of increasing the size of the depletion layer. Very few electrons have enough energy to reach the conduction band so very little current can pass.

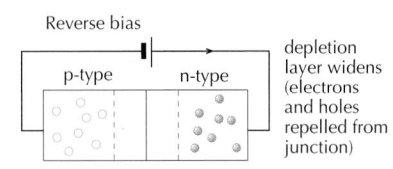

Reverse bias

p-type n-type

depletion layer widens (electrons and holes repelled from junction)

2. Forward biased

To forward bias the diode, a negative voltage is applied to the n-type side of the material and a positive voltage is applied to the p-type side of the material. This has the effect of narrowing the depletion layer and allowing the p-n junction to conduct. Electrons have enough energy to reach the conduction band, allowing current to be transferred.

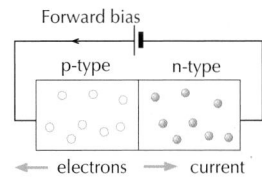

Forward bias

p-type n-type

← electrons → current

The light emitting diode (LED)

When a p-n junction is forward biased, holes combine with electrons at the junction of the diode and release energy in the process. In most semiconductors this energy is in the form of heat, however in some cases the energy is released as light. This is the basis for light emitting diodes (or LEDs).

The wavelength of the photon (colour of light) depends upon the energy released when the electron and hole combine.

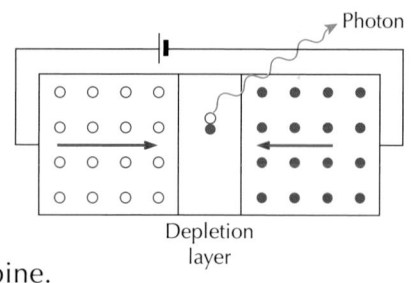

Depletion layer

Photodiodes

The photodiode is a very thin sandwich with a layer of p-type material at the top, a junction in the middle, and an n-type material at the bottom.

Light passes through the p-layer and is absorbed at the junction, where the energy of the photon produces an electron-hole pair. The electron-hole pairs recombine very quickly and produce energy.

When the photodiode uses light to produce energy, it is said to be operating in **photovoltaic** mode. This is the basis for solar cells.

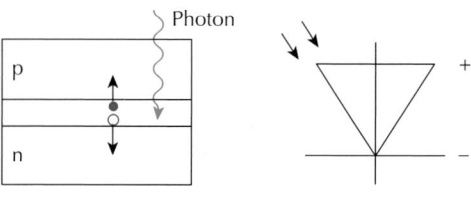

The same photodiode can be used with a reverse bias voltage applied. No current will flow unless charge carriers (electron-hole pairs) are liberated by light.

The current that is transferred is proportional to the light irradiance (see page 74). This allows the photodiode to be used as a light sensor. When operated this way, the photodiode is said to be in **photoconductive** mode. This is the basis for light dependent resistors (LDR).

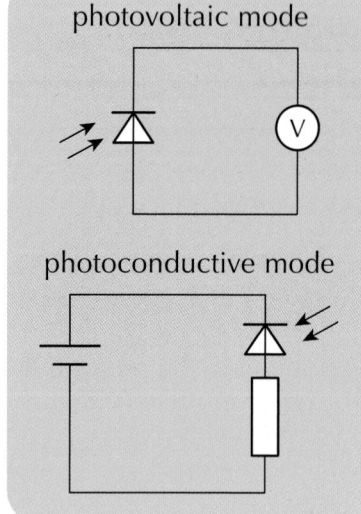

photovoltaic mode

photoconductive mode

TOP TIP

Note that in a photodiode, the photon produces an electron-hole pair. This electron-hole pair does not combine at the junction of the diode. The electron combines with holes in the p-type material and the hole combines with electrons in the n-type material.

Quick Test 51

1. When p-type and n-type materials are joined, a layer is formed at the junction. What is the name of this layer?

2. Describe how an LED emits light.

3. Describe how a diode in photovoltaic mode operates.

Quick test answers

Quick Test 1

1. A scalar quantity is described by magnitude only. A vector quantity is described by magnitude **and** direction.
2. Displacement measures how far an object moves in a given direction from its starting position.
3. Velocity is the displacement involved in a unit of time (usually seconds).
4. The object is increasing its velocity by 2 m s^{-1} every second.
5. The resultant is a vector that could replace all vectors present with a single vector.

Quick Test 2

1. Velocity
2. Displacement = area under velocity–time graph
3. Acceleration

Quick Test 3

1. The student should have written 'suvat'.
2. 27·6 m

Quick Test 4

1. An object will remain at rest, or will continue to move in a straight line at a constant speed, unless acted upon by an unbalanced force.
2. A uniform unbalanced force will produce a uniform acceleration ($F = ma$).
3. The maximum velocity an object can reach.

Quick Test 5

1. Horizontal: $F_h = F\cos\theta$, Vertical: $F_v = F\sin\theta$
2. Parallel component of weight
3. *parallel component of weight = mgsinθ*

Quick Test 6

1. Joule
2. Energy cannot be created or destroyed, only transformed from one form into another.
3. Work has to be done against gravity in order to lift an object from the ground
4. 1·85 m s^{-1}

Quick Test 7

1. $p = mv$
2. Mass and velocity
3. One momentum must be taken as being negative.

Quick Test 8

1. In the absence of external forces, the total momentum before a collision is equal to the total momentum after a collision.
2. In an elastic collision, both momentum and kinetic energy are conserved. In an inelastic collision, momentum is conserved but kinetic energy is not conserved.

Quick Test 9

1. The force of impact and the time over which the force acts.
2. Impulse = area under a force–time graph
3. Increase the time of contact, use a softer impact material
4. For every action force there is an equal and opposite reaction force *or* if object A exerts a force on object B then object B exerts an equal and opposite force on object A.

Quick Test 10

1. A projectile is an object which is influenced only by the force of gravity.
2. Horizontal and vertical
3. False; both objects hit the ground at the same time.
4. Constant velocity/zero acceleration
5. Constant acceleration

Quick Test 11

1. $v_h = v\,cos\theta$
2. $v_v = v\,sin\theta$
3. Vertical velocity is zero.
4. They are the same.

Quick Test 12

1. The horizontal launch velocity.
2. When the force of gravity is the only force acting on an object.
3. Low Earth Orbit and Geostationary Orbit.

Quick Test 13

1. The mass of the two objects and their distance apart.
2. The gravitational forces are too small.
3. The force on a 1 kg mass placed in the field.
4. Lunar and planetary orbits, satellite motion, etc.

Quick Test 14

1. Galilean invariance states that the laws of physics are the same whether you are stationary or moving at a constant speed.
2. False; in a Newtonian universe, all observers experience time in the same way.
3. (a) You; observer is moving towards you at 22 m s^{-1}.
   (b) Observer; you are moving towards the observer at 22 m s^{-1}.

Quick Test 15

1. The laws of physics are the same for all observers; the speed of light is the same for all observers.
2. Time dilation and length contraction
3. 8 hours

Quick Test 16

1. Sound, light and all the waves in the electromagnetic spectrum
2. Observed frequency decreases
3. Observed frequency increases
4. 186.3 Hz

Quick Test 17

1. Wavelength increases.
2. The galaxies are moving away from Earth.

Quick Test 18

1. The velocity at which a galaxy is moving away from us.
2. Distance and the speed of a galaxy.
3. 13·79 billion years.

Quick Test 19

1. The rate of expansion in increasing.
2. Dark energy
3. Dark matter

Answers to questions

Quick Test 20

1. Red, orange, yellow, white, blue-white
2. From the peak wavelength (colour) emitted

Quick Test 21

1. Radiation left over from the Big Bang that can be detected in all directions.
2. Hydrogen and helium
3. Light from some galaxies will never reach us; redshift of visible light to non-visible wavelengths.

Quick Test 22

1. 6 orders of magnitude (1 million = 1 000 000)
2. Particle physics
3. Distances in space

Quick Test 23

1. A particle that has the same mass as its matter counterpart, but opposite charge.
2. 12
3. Baryons consist of three quarks, mesons consist of two quarks.
4. Photon – The electromagnetic force

   Gluon – The strong nuclear force

   W and Z bosons – The weak nuclear force

Quick Test 24

1. A region where objects will experience a force
2. From positive to negative
3. Equally spaced field lines

Quick Test 25

1. Electrical potential energy/kinetic energy
2. A potential difference of 1 volt exists between two points if 1 joule of energy is required to move 1 coulomb of charge between those two points.
3. $v = 6 \cdot 2 \times 10^{5} \ m \ s^{-1}$

Quick Test 26

1. The movement of charges
2. The left-hand rule
3. The right-hand rule

Quick Test 27

1. Through rapidly changing electric fields
2. Powerful magnets/strong magnetic fields

Quick Test 28

1. A = mass number (total number of protons plus neutrons), Z = atomic number (number of protons)
2. Versions of an element that have the same atomic number, but a different mass number.
3. Reduces the mass number by 4 and the atomic number by 2
4. Increases the atomic number by 1, the mass number remains unchanged

Quick Test 29

1. Fission is the process of a heavy nucleus splitting to produce two or more lighter nuclei.
2. The mass difference between the total mass before fission and the total mass after fission
3. Control rods to absorb neutrons/coolant to transfer heat away from reactor core/containment within concrete or lead.

Quick Test 30

1. Fusion is the process of two lighter nuclei combining to form a single nucleus of larger mass.
2. The mass after a fusion reaction is less than the mass before a fusion reaction. This loss of mass is converted to energy.
3. The plasma is at such a high temperature that it would melt or evaporate any container. Strong magnetic fields have to be used to confine the plasma and keep it from contacting the edges of the container.

Quick Test 31

1. The emission of electrons when electromagnetic radiation strikes a metal surface
2. The frequency/wavelength of the electromagnetic radiation and the type of metal on which the radiation is incident
3. Photoelectrons being emitted from the metal plate
4. Light as a wave: interference, diffraction
   Light as a particle: the photoelectric effect, inside a Geiger-muller tube

Quick Test 32

1. The threshold frequency is the minimum frequency of a photon required for photoemission.

2. The work function of a metal is the minimum energy required to cause photoemission.

3. The kinetic energy of a photoelectron depends only upon the frequency or wavelength of the radiation absorbed.

Quick Test 33

1. Waves meet in phase, i.e., crest meets crest

2. Waves meet out of phase, i.e., crest meets trough

3. The two wave sources must have a constant phase difference and have the same wavelength, frequency and velocity.

4. The extra distance travelled by one wave source compared to another wave source.

Quick Test 34

1. For a maxima, the path difference is a whole number of wavelengths; for a minima, the path difference is an odd number of half wavelengths, e.g., 1.5λ or 4.5λ.

2. The distance, d, between adjacent slits; the wavelength λ; the distance, D, between the slits and the screen

3. 100 lines per mm = 100 000 lines per metre, $d = \dfrac{1}{100\,000} = 1 \times 10^{-5}\ m$

4. $13.9°$

Quick Test 35

1. Refraction is the change in direction when a wave moves from one medium to another.

2. The absolute refractive index is the refractive index of a material compared to the refractive index of a vacuum.

3. The ratio of sine of the angles in the two media; the ratio of the speeds in the two media; the ratio of the wavelengths in the two media.

4. $1.8 \times 10^{8}\ m\ s^{-1}$

Quick Test 36

1. The critical angle is the angle of incidence for a material that results in an angle of refraction of 90 degrees.

2. The light is totally internally reflected.

Quick Test 37

1. Irradiance is the power of radiation per unit area.

2. Irradiance varies with the square of the distance, i.e., if the distance doubles, the irradiance reduces by a factor of 4.

Quick Test 38

1. A continuous spectrum is a spectrum of light with no visible separate lines.

2. As light from hotter regions of the Sun passes through cooler regions of the upper atmosphere, atoms absorb light at specific frequencies/wavelengths to produce an absorption spectrum. These absorption lines allow the elements present to be identified.

Quick Test 39

1. The ground state is the lowest energy level in the Bohr model of the atom.

2. The ionisation level is where the electron escapes the atom and reaches a point of zero energy.

3. By absorbing a photon of electromagnetic radiation that has energy equal to precisely the difference between the energy levels.

4. The electron loses energy as it moves to a lower energy level and emits a photon in the process. The energy of the photon is equal to the energy difference between levels.

Quick Test 40

1. Take reading of timebase from oscilloscope. Count number of divisions for one wave (period) and multiply this number by the figure from the timebase dial. Calculate frequency from 1/period.

2. $V_{peak} = \sqrt{2}\, V_{rms}$

   $12 = \sqrt{2}\, V_{rms}$

   $V_{rms} = 8 \cdot 5\, V$

3. Some examples: DC cannot be transmitted across large distances without power losses. AC can produce very high voltages.

4. Double the number of waves will appear on the screen.

Quick Test 41

1. Voltage in series circuits is split up across the components of the circuit; voltage is the same across every 'branch' of a parallel circuit.

2. Current in series circuits is the same at all points; Current in parallel circuits is split up in the branches of the parallel circuit

3. $P = \dfrac{E}{t}$ $P = IV$ $P = I^2 R$ $P = \dfrac{V^2}{R}$

Quick Test 42

1. The potential difference across the voltmeter is 0 V.

2. Altering the values of R_1, R_2, R_3 or R_4 would put the bridge out of balance.
   If the supply voltage was increased or decreased then the bridge would still be

   balanced as $\dfrac{R_1}{R_2} = \dfrac{R_3}{R_4}$

Quick Test 43

1. The energy supplied to each coulomb of charge passing through a power supply when no current is drawn from the supply.

2. The terminal potential difference is the energy per coulomb of charge available to an external circuit when current is drawn from the cell/battery.

3. The internal resistance of the supply causes energy to be lost due to collisions between charged particles and other atoms within the power supply.

Quick Test 44

1. The y-intercept gives the EMF and the gradient of the line gives negative internal resistance.

2. By connecting the terminals of a power supply together with a short length of wire. This is called the short-circuit current.

Quick Test 45

1. Capacitance is the charge per volt.

2. $C = \dfrac{Q}{V}$ or $Q = CV$

3. Capacitance is the gradient of the line in a Q against V graph.

Quick Test 46

1. Energy has to be supplied by the battery in order to overcome the repulsion between the negative charges and place more negative charge on one plate of the capacitor, or to overcome the attraction between the positive and negative charges and remove negative charge from the other plate of the capacitor.

2. The energy stored in capacitor = area under QV graph = $\dfrac{1}{2}QV$.

Quick Test 47

1. Increasing capacitance increases the time to charge and discharge a capacitor. Decreasing capacitance reduces the time it takes to charge and discharge a capacitor.

2. An increase in resistance in an RC circuit will decrease the maximum value of current, as well as increasing the time for the capacitor to charge and discharge.

Quick Test 48

1. Provides a very high current for a short period of time, resulting in rapid transfer of charge – creating a 'flash'.

2. Noise suppression/AC to DC conversion/capacitive touchscreens.

Quick Test 49

1. Conductors, insulators and semiconductors

2. The valence band is the highest energy band that electrons will occupy at room temperature.

3. The conduction band is the highest energy band in a material. The electrons are free to move in the conduction band.

4. The energy gap between the valence band and the conduction band is far too large for an insulator to conduct at room temperature. The energy gap between the valence band and the conduction band in a semiconductor is much smaller, allowing electrons to move into the conduction band and current to flow.

Quick Test 50

1. Doping is the process of adding impurities to a semiconductor in order to change its conductivity.

2. Atoms with five outer electrons are added to a silicon crystal lattice to produce free electrons.

3. Atoms with three outer electrons are added to a silicon crystal lattice to produce positive holes.

4. Doping semiconductors reduces their resistance, increasing the conductivity.

Quick Test 51

1. The depletion layer

2. A forward biased diode causes electrons and holes to combine at the junction of the diode and release energy in the form of a photon of light.

3. A photon is absorbed at the junction of the diode which produces an electron-hole pair. The electron and hole recombine to produce energy.